Digital Signal Processing for Complete Idiots

DAVID SMITH

ISBN: 9781086340358

DEDICATION

This book is dedicated to all scientists, who work to make the world a better place.

CONTENTS

PREFACE

Digital Signal Processing (DSP) is a subject of central importance in engineering and the applied sciences. Signals are information-bearing functions, and DSP deals with the analysis and processing of signals (by dedicated systems) to extract or modify information. Signal processing is necessary because signals normally contain information that is not readily usable or understandable, or which might be disturbed by unwanted sources such as noise. Although many signals are nonelectrical, it is common to convert them into electrical signals for processing. Most natural signals (such as acoustic and biomedical signals) are continuous functions of time, also referred to as analog signals. Prior to the development of DSP, Analog Signal Processing (ASP) and analog systems were the only tools to deal with analog signals.

Although analog systems are still widely used, Digital Signal Processing (DSP) and digital systems are attracting more attention, due in large part to the significant advantages of digital systems over their analog counterparts. These advantages include superiority in performance, speed, reliability, efficiency of storage, size and cost. In addition, DSP can solve problems that cannot be solved using ASP, like the spectral analysis of multi-component signals, adaptive filtering, and operations at very low frequencies. Following the recent developments in engineering which occurred in the 1980s and 1990s, DSP became one of the world's fastest growing industries. Since that time DSP has not only impacted on traditional areas of electrical engineering, but has had far reaching effects on other domains that deal with information such as economics, meteorology, seismology, bioengineering, oceanology, communications, astronomy, radar engineering, control engineering and various other applications.

DSP is a very math intensive subject and one would require a deep understanding in mathematics to understand various aspects of DSP. We believe to explain science with mathematics takes skill, but to explain science without mathematics takes even more skill. Although there are many books which cover DSP, most of them or all of them would require a ton of mathematics to understand even the most fundamental concepts. For a first timer in DSP, getting their heads around advanced math topics like Fourier transform etc. is a very hard task. Most students tend to lose interest in DSP, because of this sole reason. Students don't stick around long enough to discover how beautiful a subject DSP is.

In this book, we've tried to explain the various fundamental concepts of DSP in an intuitive manner with minimum math. Also, we've tried to connect the various topics with real life situations wherever possible. This way even first timers can learn the basics of DSP with minimum effort. Hopefully the students will enjoy this different approach to DSP. The various concepts of the subject are arranged logically and explained in a simple reader-friendly language with MATLAB examples.

Finally, this books is not meant to be a replacement for those standard DSP textbooks, rather this book should be viewed as an introductory text for beginners to come in grips with advanced level topics covered in those books. This book will hopefully serve as inspiration to learn DSP in greater depths.

Readers are welcome to give constructive suggestions for the improvement of the book.

Thank you!

1. SIGNALS

1.1 SIGNALS

We use the term 'signal' a lot in our daily life, and our idea of a signal is some kind of wave that transmits information. This is a fairly accurate interpretation of the term signal. But what really is a signal? A signal is a mathematical representation of a function of one or more independent variable. A signal describes how one parameter varies with another. For example, the variation of temperature of your room with respect to time is a signal. Voltage changing over time in an electrical circuit is also a signal. In DSP, the independent quantity we are dealing with is time.

There are two basic types of signals; Continuous time signals and Discrete time signals. Continuous time signals are those signals that are defined for every instant of time. Discrete time signals on the other hand are those signals whose values are defined only at certain instants of time. For example, if you take the temperature reading of your room after every hour and plot it, what you get is a discrete time signal. The temperature values are only defined at the hour marks and not for the entire duration of time. The value of temperature at other instants (say at half or quarter hour marks) are simply not defined.

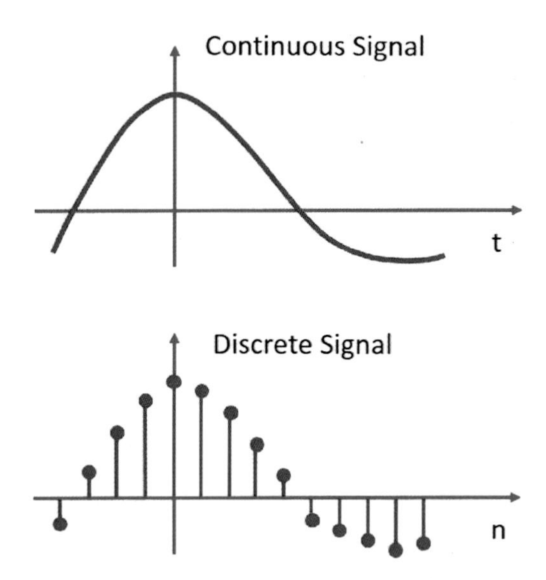

For Continuous time signals, the independent variable is represented as **t** (continuous time) and for Discrete time signals, the independent variable is represented as **n** (equally spaced instants of time). The dependent variables are represented as **x(t)** and **x[n]** respectively.

1.2 BASIC CONTINUOUS TIME SIGNALS

In this section, we introduce several important continuous time signals. The proper understanding of these signals and their behavior will go a long way in making DSP an easier subject.

1.2.1 Sinusoids

Sine waves and Cosines waves are collectively called sinusoids or sinusoidal signals. Mathematically, they are defined as:

$$x(t) = A\cos(\omega t + \Phi)$$

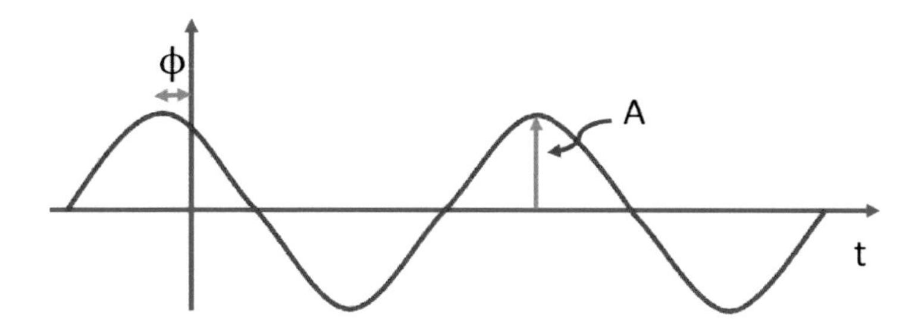

Where, A is the Amplitude (maximum value/height of the signal), ω is the angular frequency and ϕ is the phase.

It is important to understand that both Sine waves and Cosines waves are basically the same, except that they start at different times i.e. they are 90 degrees out of phase. Time period of a sinusoid is $T = 2\Omega / \omega$. Sine and cosine waves of same frequency can be represented as a single entity using complex notation with the help of Euler's formula as,

$$e^{i\omega t} = \cos(\omega t) + i \sin(\omega t)$$

This representation makes calculations a lot easier (although it may not seem so at first glance) and is used extensively throughout this book.

1.2.2 Unit Step Signal

A Unit Step Signal is mathematically defined as:

$$\begin{aligned} u(t) &= 0, \qquad t > 0 \\ &= 1, \qquad t < 0 \end{aligned}$$

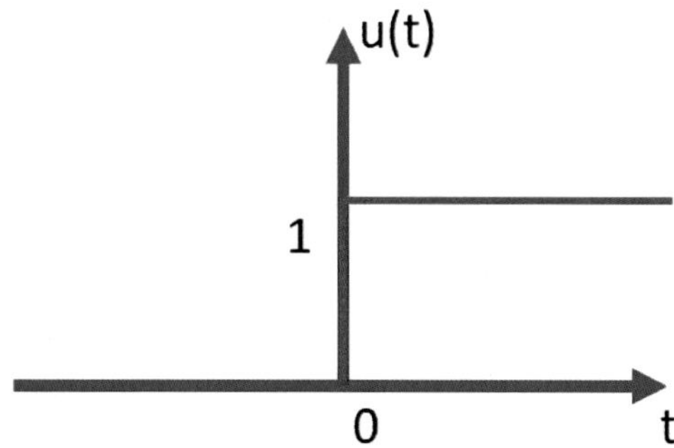

The step signal can be imagined as the output of a bulb being turned on at t = 0. Initially it has zero output, then after it is turned on it has a constant output for the rest of the time. The Unit step signal is discontinuous at t = 0, but for the sake of simplicity you can take u(0) =1. An important thing to be noted here is that, a Continuous signal may not be a continuous function mathematically, as evident from this case.

The Unit step signal is of immense importance in engineering, it is used to study the steady state performance of systems. Sometimes you may encounter step signals of different magnitudes, they are basically the scaled version of the Unit step signal

1.2.3 Unit Impulse Signal

Another very important basic signal is the Unit impulse signal. It is mathematically defined as:

$$\delta(t) = 0, \quad t \neq 0$$
$$= \infty, \quad t = 0$$

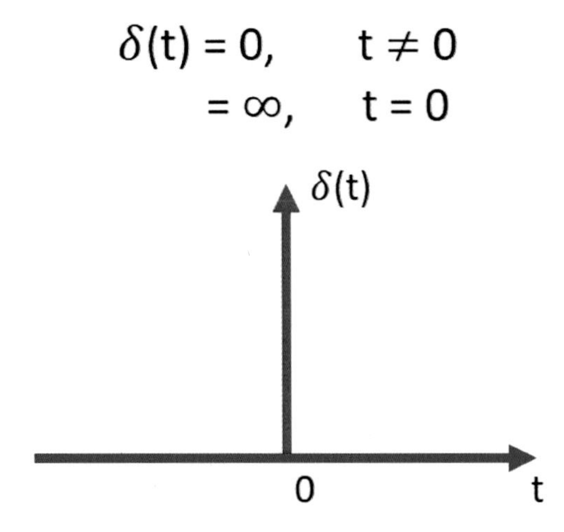

You can think of the Impulse signal as a short pulse, like the output when a bulb is switched on and off as fast as you can. The unit Impulse function is also known as the delta function or the delta-dirac function and is represented as $\boldsymbol{\delta(t)}$.

A doubt you may have is, if the value of the Unit Impulse function is ∞ at t = 0, then why the name Unit Impulse function? The name comes from the fact that the Unit impulse function has a unit area at t = 0.

Consider a rectangle of width ε and height 1/ε as shown in the figure. The area of the rectangle is unity. Now if you make ε infinitesimally small keeping the area unity, what you get is the unit impulse function.

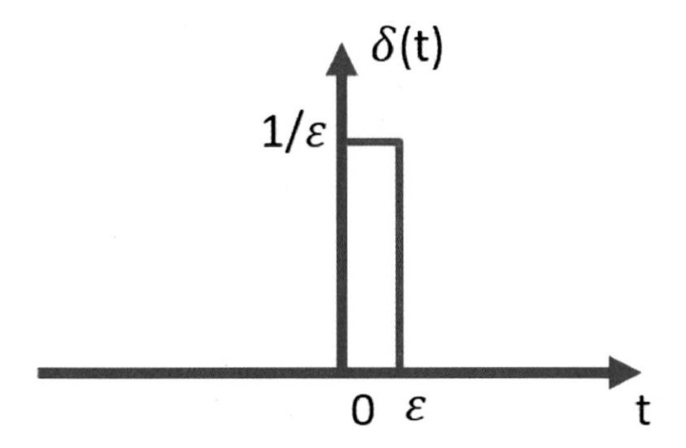

The height of the arrow is used to depict the area of a scaled impulse.

The Unit step and the Unit impulse signal are related to each other as:

$$\frac{d\,u(t)}{dt} = \delta(t)$$

This relation is self-explanatory.

1. for $t < 0$, $u(t) = 0$, therefore slope = 0
2. for $t > 0$, $u(t) = 1$, therefore slope = 0
3. at $t = 0$, $u(t)$ changes from 0 to 1, therefore the slope = ∞

The relationship can be rewritten in another form as:

$$\int_{-\infty}^{t} \delta(\tau)\, d\tau = u(t)$$

i.e. a Unit step signal is the running integral of the Unit impulse signal.

1.2.4 Exponential Signal

An exponential signal is that signal which rises or decays exponentially (by the power of e). It is mathematically defined as:

$$x(t) = C\, e^{at}$$

Where **e** is the Euler's number (**e = 2.71828**), **C** and **a** are constants.

The characteristics of an exponential signal depends upon the values of **C** and **a**. These values maybe real or complex no.'s, and as mentioned earlier, an exponential function with complex constant **a** is basically a sinusoid.

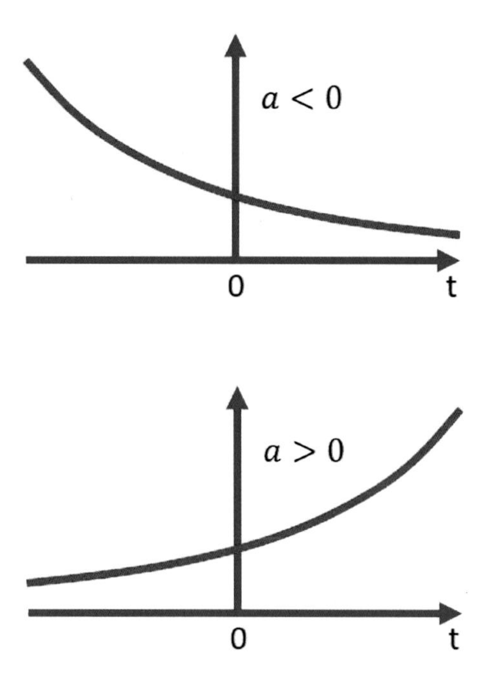

There are other basic signals too, like the ramp signal, triangular signal etc. But in DSP we are mostly dealing with Impulse signals and Step signals.

1.3 BASIC DISCRETE TIME SIGNALS

All the basic signals discussed in the last section have a discrete version too.

1.3.1 Discrete Sinusoids

A discrete sinusoid is mathematically defined as:

$$x[n] = A\cos(\Omega n + \phi)$$

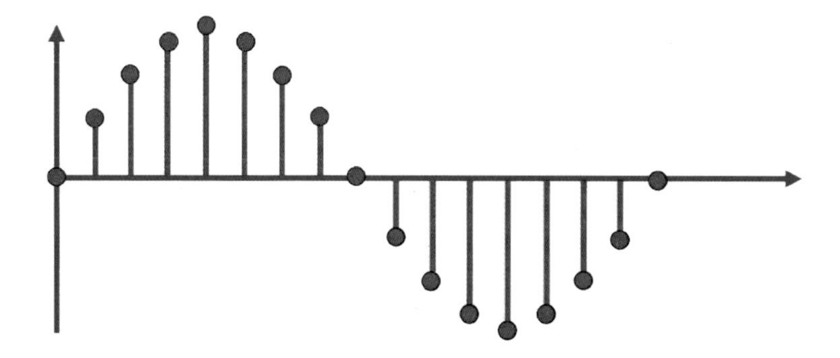

All the properties of Discrete Sinusoids are the same as their continuous counterparts.

1.3.2 Discrete Unit step signal

Discrete time Unit step signal is defined as:

$$u[n] = \begin{cases} 0, & n < 0 \\ 1, & n \geq 0 \end{cases}$$

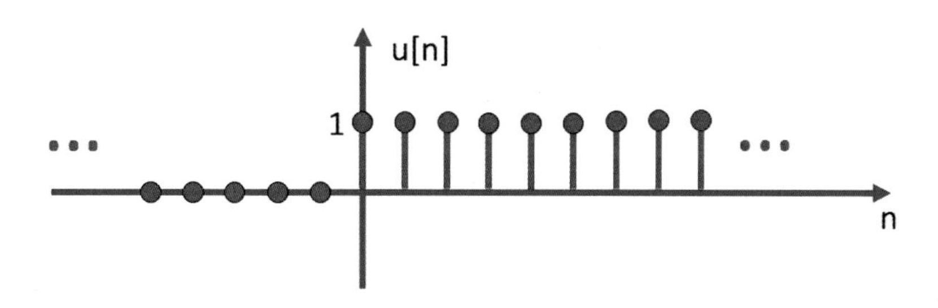

The value of discrete step signal is unity at n = 0.

1.3.3 Discrete Unit Impulse signal

Discrete time Unit impulse signal is defined as:

$$\delta[n] = \begin{cases} 0, & n \neq 0 \\ 1, & n = 0 \end{cases}$$

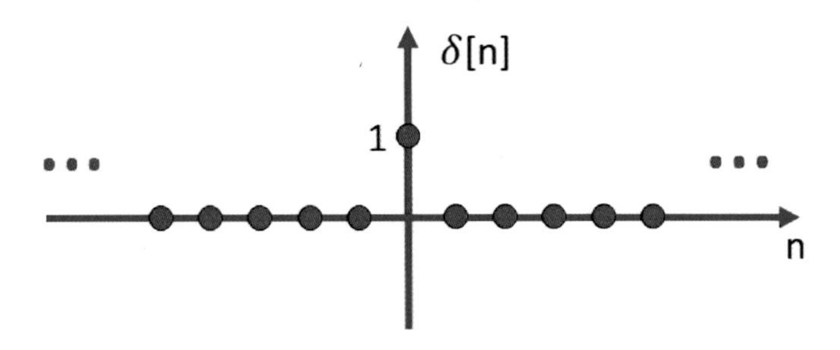

Unlike in the case of Continuous Unit impulse signal, the Discrete impulse signal has a fixed magnitude at n = 0.

1.3.4 Discrete Exponential signal

Discrete time Exponential signal is defined as:

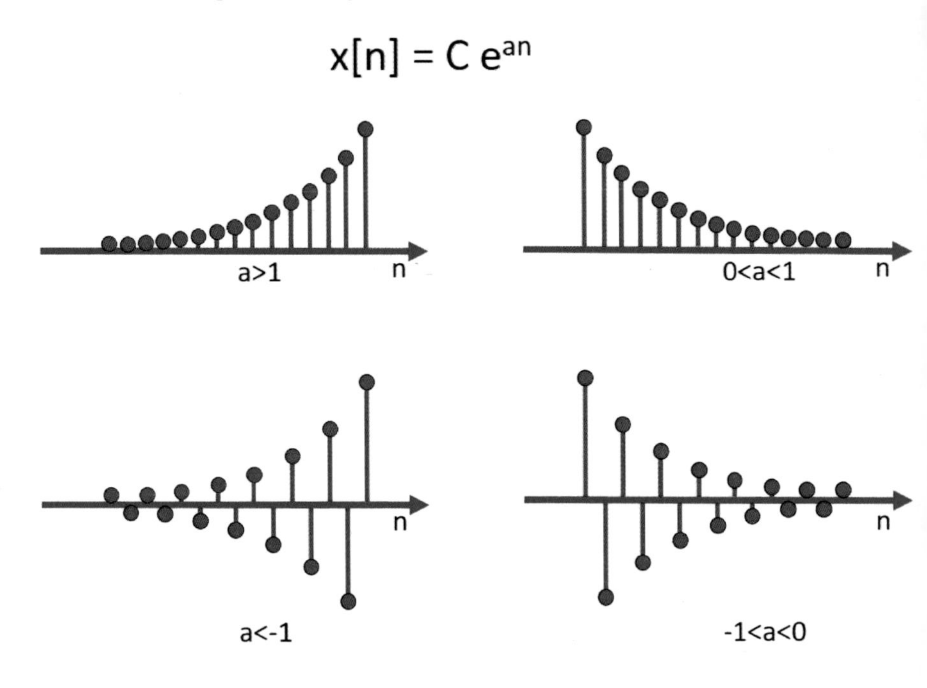

1.4 BASIC SIGNAL OPERATIONS

There are 2 variable parameters in a signal: Amplitude and Time. By varying these parameters, we can perform some basic operations on signals.

1.4.1 Amplitude Scaling

Amplitude scaling is nothing but multiplying the amplitude by a scalar (real) quantity. If the scalar quantity is greater than one, then the resultant signal is amplified and the process is called Amplification. If the scalar quantity is less than one, then the resultant signal is diminished and the process is called Attenuation. Amplitude scaling is always done for the signal as a whole, and never just a portion of the signal.

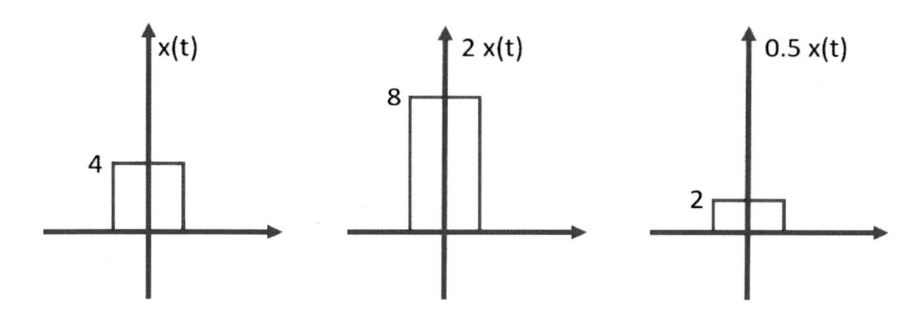

Amplitude scaling is denoted as: **y(t) = a x(t)**, where **a** is the scaling factor.

1.4.2 Addition

Two signals can be added by adding their corresponding amplitudes at the same instants of time.

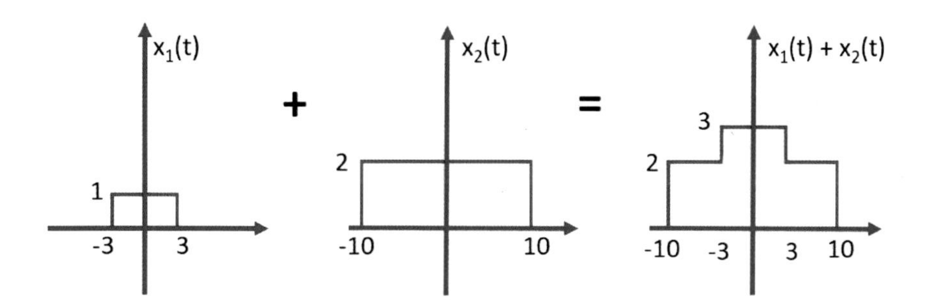

The addition operation is denoted as: $\mathbf{y(t) = x_1(t) + x_2(t)}$

In the figure above,

- $-10 < t < -3$, $z(t) = x_1(t) + x_2(t) = 0 + 2 = 2$
- $-3 < t < 3$, $z(t) = x_1(t) + x_2(t) = 1 + 2 = 3$
- $3 < t < 10$, $z(t) = x_1(t) + x_2(t) = 0 + 2 = 2$

1.4.3 Multiplication

Multiplication of two signals can be performed by multiplying their corresponding amplitudes at the same instants of time.

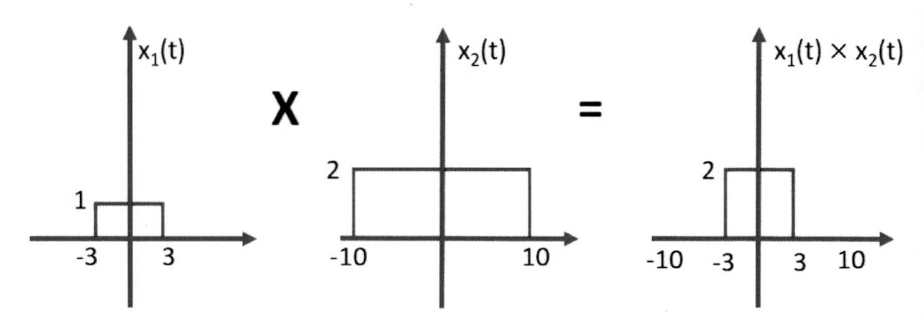

The multiplication operation is denoted as: $\mathbf{y(t) = x_1(t) \times x_2(t)}$

In the figure above,

- $\mathbf{-10 < t < -3,\ z\ (t) = x_1(t) \times x_2(t) = 0 \times 2 = 0}$
- $\mathbf{-3 < t < 3,\ z\ (t) = x_1(t) \times x_2(t) = 1 \times 2 = 2}$
- $\mathbf{3 < t < 10,\ z\ (t) = x_1(t) \times x_2(t) = 0 \times 2 = 0}$

1.4.4 Time shifting

Time shifting of a signal simply means to shift the starting instant of a signal to an earlier or a later time or in other words, time shifting means fast-forwarding or delaying a signal.

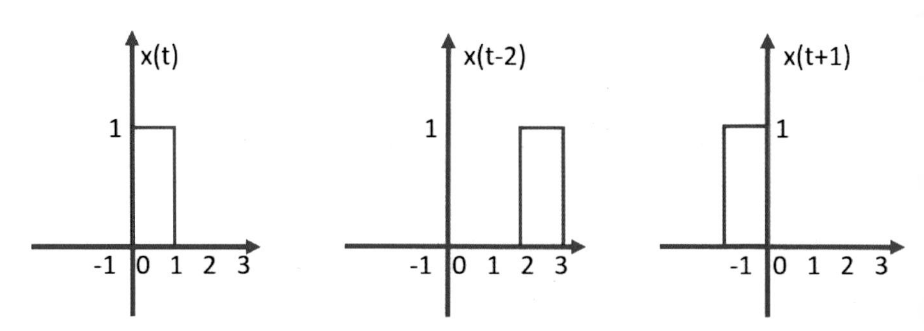

The time shifting operation is denoted as: $\mathbf{y(t) = x(t-t_0)}$

In the figure above, the second signal is y(t) = x(t - 2), this is nothing but the signal x(t) that starts 2 seconds later i.e. it is delayed by 2 seconds. The third signal is y(t) = x(t + 1), this is the signal x(t) that starts 1 second earlier i.e. it is fast forwarded by 1 second.

1.4.5 Time scaling

Time scaling of signals involves the modification of periodicity of the signal, keeping its amplitude constant. In laymen terms, time scaling means either expanding or compressing a signal without changing its amplitude.

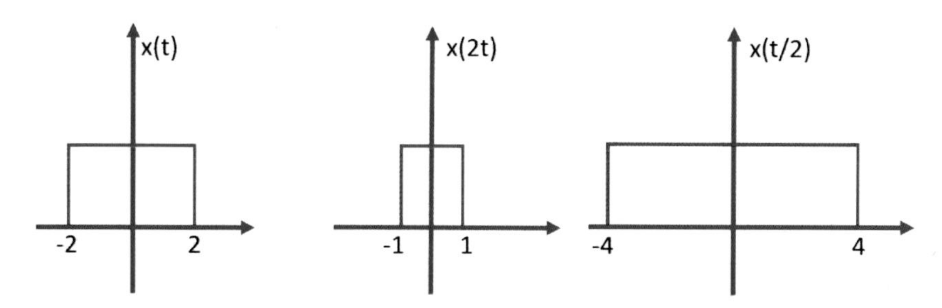

Have you ever played a song at twice the speed on your music player? Have you wondered how it's possible? It is possible because of time scaling, time compressing to be exact. In doing so the song isn't distorted in any way, the words, the instruments are all there and even the loudness doesn't change. That's because time scaling has nothing to do with the amplitude.

The time scaling operation is denoted as: **y(t) = x(at)**, where **a** is a constant. When a > 1, the signal is compressed and when a < 1, the signal is expanded. For instance, the signal y(t) = x(2t) is the compressed to half version of x(t). Seems counterintuitive? Suppose you plot a graph (any figure) with the scale 10 divisions = 10 units on the x-axis. Then if you plot the same graph with the scale 10 divisions = 20 units, what difference do you see? The plot is the same, it just got compressed by half. This is exactly what happens to a signal. By changing the independent variable **t** to **at**, we are essentially forcing the signal to be **a** times faster/slower, therefore the signal duration changes accordingly.

Do note that it is not possible to time scale an impulse function.

Although we have explained the signal operations using continuous time signals, they are applicable in exactly the same manner for Discrete time signals.

1.5 MATLAB

1.5.1 Basic signals (Discrete)

```
n = 4;
t= -n:1:n;
y1=[zeros(1,n), 1, zeros(1,n)];
subplot(2,2,1);
stem(t,y1);
ylabel('d[n]');
xlabel('n');
title('Unit impulse function');

n = 4;
t =0:1:n;
y2=[ones(1,n+1)]
subplot(2,2,2);
stem(t,y2);
ylabel('u[n]');
xlabel('n');
title('Unit step function');

n = 4;
t =0:1:n;
subplot(2,2,3);
stem(t,t);
ylabel('r[n]');
xlabel('n');
title('Unit ramp function');

n = 4;
a=2;
t =0:1:n;
y3=exp(a*t);
subplot(2,2,4);
stem(t,y3);
ylabel('x[n]');
xlabel('n');
title('Exponential function');
```

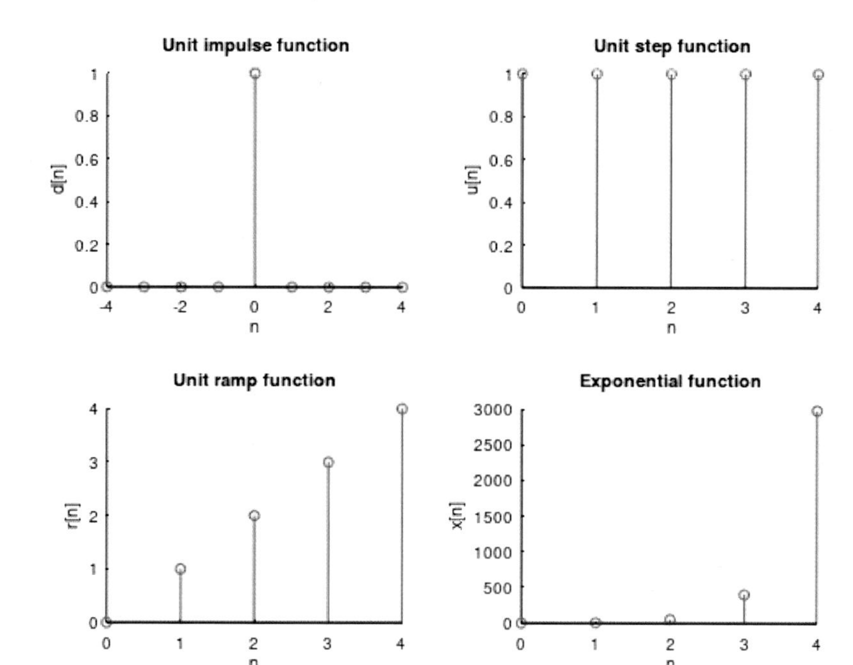

1.5.2 Basic signals (Continuous)

```
n = 4;
t= -n:0.1:n;
impulse = t==0;
subplot(2,2,1);
plot(t,impulse,'r');
ylabel('d(t)');
xlabel('t');
title('Unit impulse function');

n = 4;
t =-2:0.1:n;
unitstep = t>=0;
subplot(2,2,2);
plot(t,unitstep,'r');
ylabel('u(t)');
xlabel('t');
title('Unit step function');
```

```
n = 4;
t =-2:0.1:n;
ramp = t.*unitstep;
subplot(2,2,3);
plot(t,ramp,'r');
ylabel('r(t)');
xlabel('t');
title('Unit ramp function');

n = 4;
a=2;
t =-2:0.1:n;
y = exp(a*t);
subplot(2,2,4);
plot(t,y,'r');
ylabel('x(t)');
xlabel('t');
title('Exponential function');
```

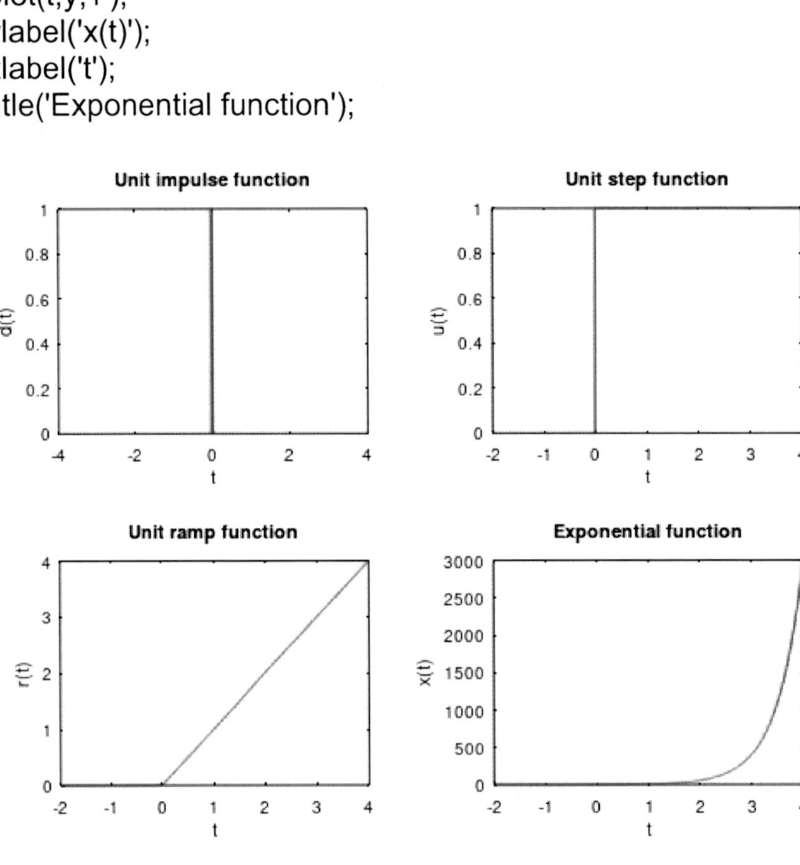

1.5.3 Sine and Cosine Signal (Discrete)

```
n =6;
t = 0:0.2:n;
subplot(1,2,1);
y = sin(t);
stem(t,y)
ylabel('Amplitude');
xlabel('n');
title('Sine function');

n =6;
t = 0:0.2:n;
subplot(1,2,2);
y = cos(t);
stem(t,y)
ylabel('Amplitude');
xlabel('n');
title('Cos function');
```

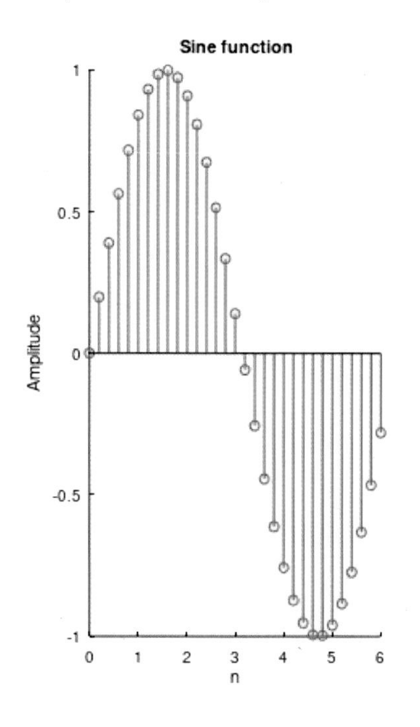

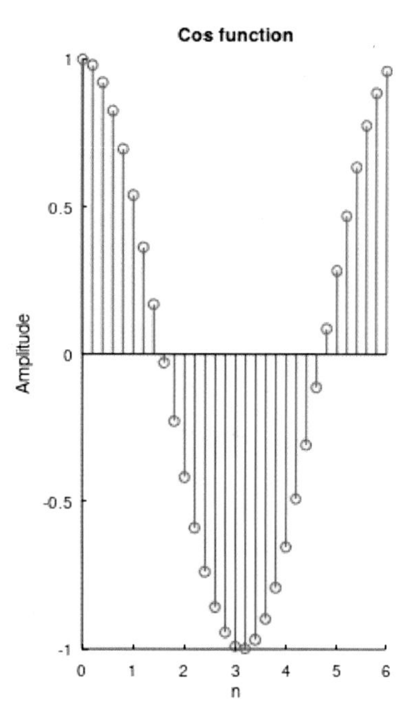

1.5.4 Sine and Cosine Signal (Continuous)

```
n =12;
t = 0:0.2:n;
subplot(1,2,1);
y = sin(t);
plot(t,y,'r')
ylabel('Amplitude');
xlabel('t');
title('Sine function');

t = 0:0.2:n;
subplot(1,2,2);
y = cos(t);
plot(t,y,'r')
ylabel('Amplitude');
xlabel('t');
title('Cos function');
```

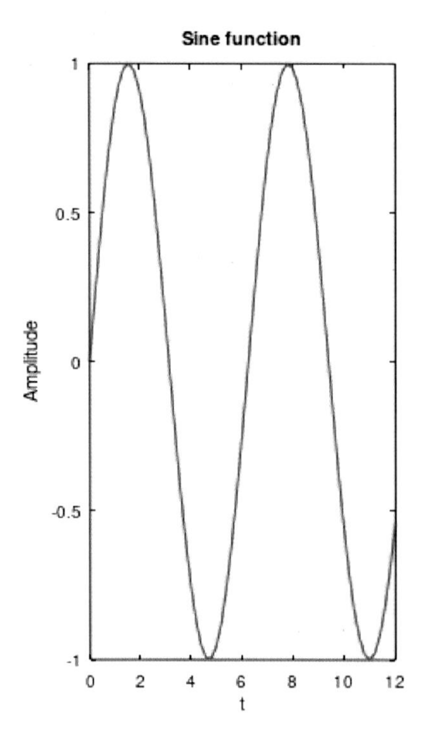

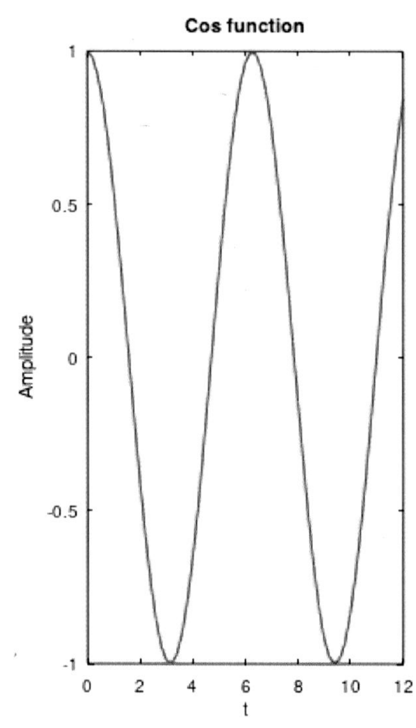

2. SYSTEMS

2.1 SYSTEMS

A system is any process or a combination of processes that takes signals as the input and produces signals as the output. For example, an amplifier that takes in an audio input and produces its amplified version is a system.

Systems that take continuous time signal inputs and produces continuous time signal outputs are called Continuous Time systems.

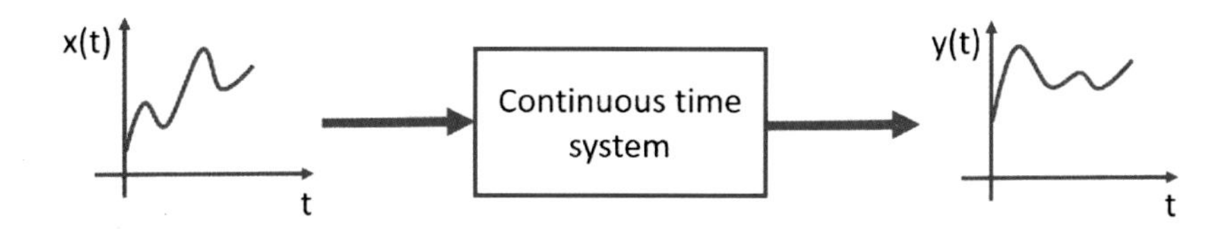

Similarly, Discrete time systems are those that takes Discrete time signal inputs and produces Discrete time signal outputs.

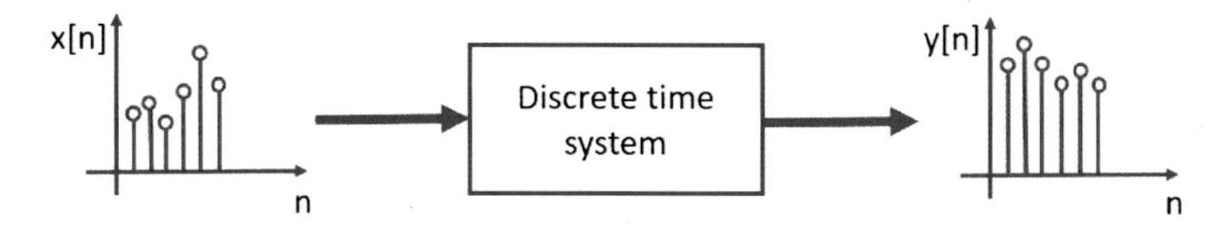

Besides these two types there are some systems that takes in continuous inputs and produces discrete outputs and vice versa. We will discuss those in another chapter.

2.2 INTERCONNECTION OF SYSTEMS

Engineers often connect many smaller systems called sub-systems together to form a new system. One big advantage of doing things this way is that it's easier to model and analyze smaller systems than large ones.

Some common types of interconnections are listed below:

2.2.1 Series or Cascade Connection

Series (or cascade) connection is the simplest type of system interconnection. In this connection, a system's output is fed as the input to another system.

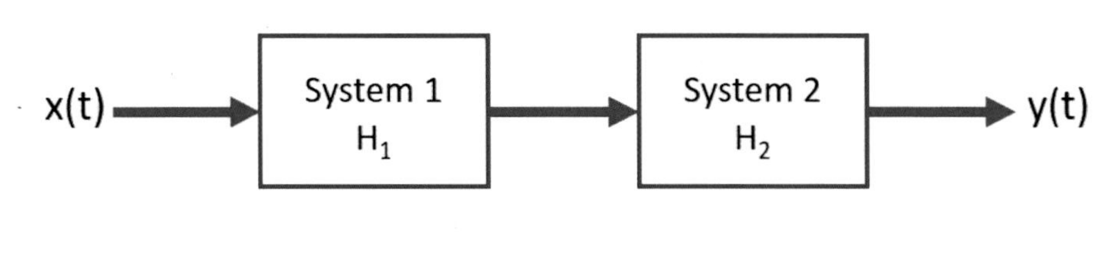

$$y(t) = H_1 (x(t)) * H_2 (x(t))$$

2.2.2 Parallel Connection

In a Parallel connection, the same input is fed to two or more systems and the corresponding outputs are summed at the end.

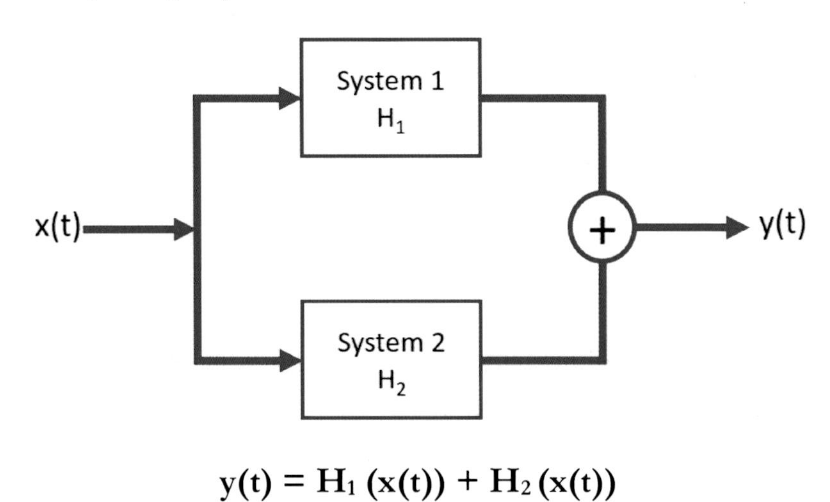

$$y(t) = H_1 (x(t)) + H_2 (x(t))$$

2.2.3 Feedback Connection

In the previous two interconnections, the system's present output has no bearing on the future outputs of the system. Whereas in a feedback connection, a portion of the previous output is fed back to the system along with the current input. The output is usually passed through another system H_2 before being fed back for comparison with the input. This is self-correcting mechanism, when the output of the system diverts from the desired output, an error signal, which is the difference between the input and the fed back output is generated. This ensures that the output rises/falls back to the desired range automatically.

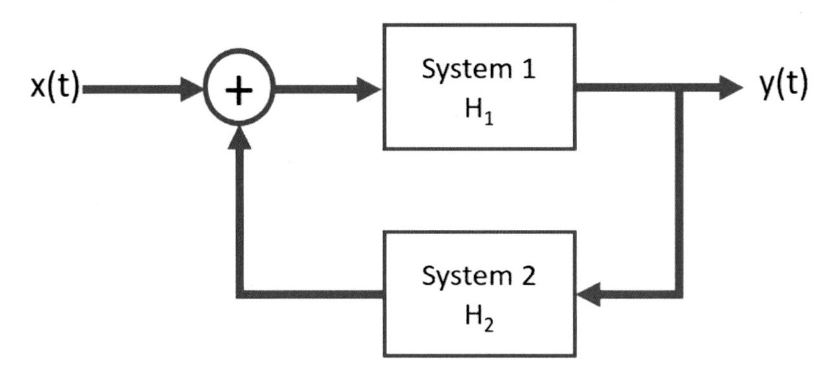

What we have defined is a negative feedback. There is also a positive feedback connection, in which the fed back previous output is summed with the current input, to drive the system.

$$\mathbf{y(t) = H_1 (x(t)) + H_2(y(t))}, \text{ for positive Feedback}$$

$$\mathbf{y(t) = H_1 (x(t)) - H_2(y(t))}, \text{ for negative Feedback}$$

2.3 PROPERTIES OF SYSTEM

In this section we introduce a couple of basic continuous and discrete time system properties.

2.3.1 Memory of the System

A system is said to be memoryless if its output at any instant depends only on the input at the same instant i.e. memoryless systems are independent of inputs and outputs in the past or the future.

Example: **y(t) = x(t-1)**

In this system, the output of the system at t=2, y(2) depends on input at time t=1, x(2-1) = x(1). So this system has a memory.

Feedback systems discussed in the previous section are an example of systems with memory, since their current output depends not only on present input, but also past outputs.

2.3.2 Causality

A system is said to be causal if the output at any instant depends only on the values of the input at the present instant or past instant. In other words, a causal system does not anticipate the future values of the input.

Example: **y(t) = x(t+1)**

In this system, the output of the system at t=2, y(2) depends on the input at time t=3, x(2+1) = x(3). So this system is a non-causal system.

All real time physical systems are causal, because time only moves forward, effect occurs after cause.

2.3.3 Time Invariance

A system is said to be time invariant, if the system behavior doesn't vary with time. Such systems are expected to produce the same outputs for the same inputs applied at a different time. This implies that for time invariant systems, a shift in the input signal should produce the same shift in the output signal.

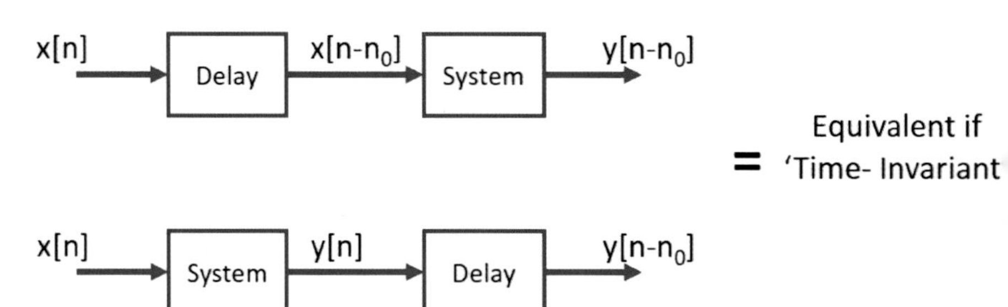

Example: **y[n] = x[n]**

This is a time invariant system, since the input delayed by 1 second produces an output delayed by 1 second.

On the other hand, systems whose behavior varies with time are called Time varying systems. The output of such systems not only depends on the input, but also the time at which the input is applied.

Example: $\mathbf{y(t) = t\, x(t)}$

This is a time variant system. It can be easily verified; if you delay the input by 1 second and then pass it to the system, the output will be $\mathbf{t\, x(t-1)}$. Instead, if you pass the input to the system first and then delay the signal, the output will be $\mathbf{(t-1)\, x(t-1)}$.

2.3.4 Stability

A stable system is one in which small inputs does not lead to erratic responses that doesn't grow out of control. Consider 2 practical cases (shown in the figure) to get some clarity on the concept of stability.

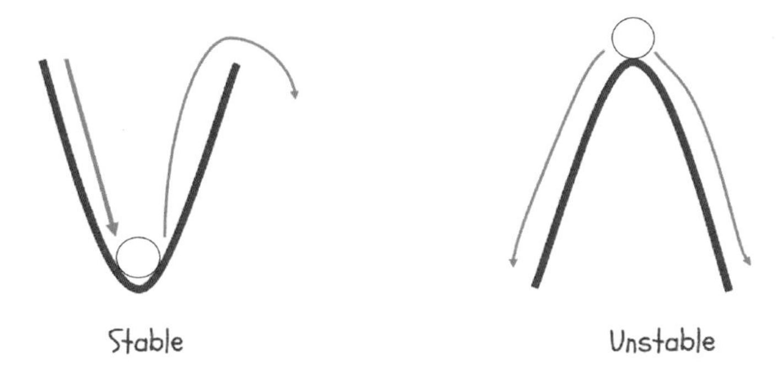

In the first case, a ball is placed in a u-shaped container. If we apply a small force on the ball, it moves back and forth slightly, but ultimately returns to its original position. Such a system is called a Stable system. In the second example, the ball is placed on top on a hill, with slope on either side. In this case, the smallest disturbance will make the ball roll down the hill. Such a system is called an Unstable system.

In DSP, a concept called '**BIBO**' is used to define the stability of a system. BIBO stands for Bounded Input Bounded Output. A bounded signal is any signal which in which the absolute value of the signal is never greater than some value. So a system is said to be BIBO stable, if we input a signal with absolute value less than some constant, and we get an output signal with absolute value less than some other constant. Since this constant value is arbitrary, we can consider it as infinity i.e. the output of a BIBO stable system should at no point tend to infinity, including the end behavior, for any finite input. The Unit impulse is usually the bounded signal of choice to determine the stability of a system.

Example: $\mathbf{y(t) = t\, x(t)}$

Suppose we provide a bounded input, say $\mathbf{x(t) = 2}$, to the system, the corresponding output will be $\mathbf{y(t) = 2t}$. The output here is not bounded, it tends to infinity as time tends to infinity. Thus this is an Unbounded system.

2.3.5 Linearity

Linearity is perhaps the most important system property. A Linear system is a system that obeys the Superposition property.

The Superposition property is a basically a combination of 2 system properties:

1. Additivity property:

A system is said to be additive if the response of a system when 2 or more inputs are applied together is equal to the sum of responses when the signals are applied individually. Suppose we provide an input (x_1) to a system and we measure its response. Next, we provide a second input (x_2) and measure its response. Then, we provide the two inputs ($x_1 + x_2$) at once. If the system is linear, then the measured response will be just the sum of the responses we noted down while providing the two inputs separately.

i.e. if $\mathbf{x_1(t) \rightarrow y_1(t)}$ **and** $\mathbf{x_2(t) \rightarrow y_2(t)}$,
Then the system is additive, if $\mathbf{x_1(t) + x_2(t) \rightarrow y_1(t) + y_2(t)}$

2. Homogeneity or Scaling property:

A system is said to obey Homogeneity property, if the response of a system to a scaled input is the scaled version of the response to the unscaled input. This means that, as we increase the strength of an input signal to a linear system, then the output strength will also be increased proportionally. For example, the human ear is a linear system, because if a person speaks twice as loud, the ear also responds twice as much.

i.e. if $\mathbf{x(t) \rightarrow y(t)}$,
Then the system obeys homogeneity, if $\mathbf{a\, x(t) \rightarrow a\, y(t)}$, where $\mathbf{a}$ is constant.

Combining both these properties we get the superposition property.

$$\mathbf{a\,x_1(t) + b\,x_2(t) \rightarrow a\,y_1(t) + b\,y_2(t)}$$

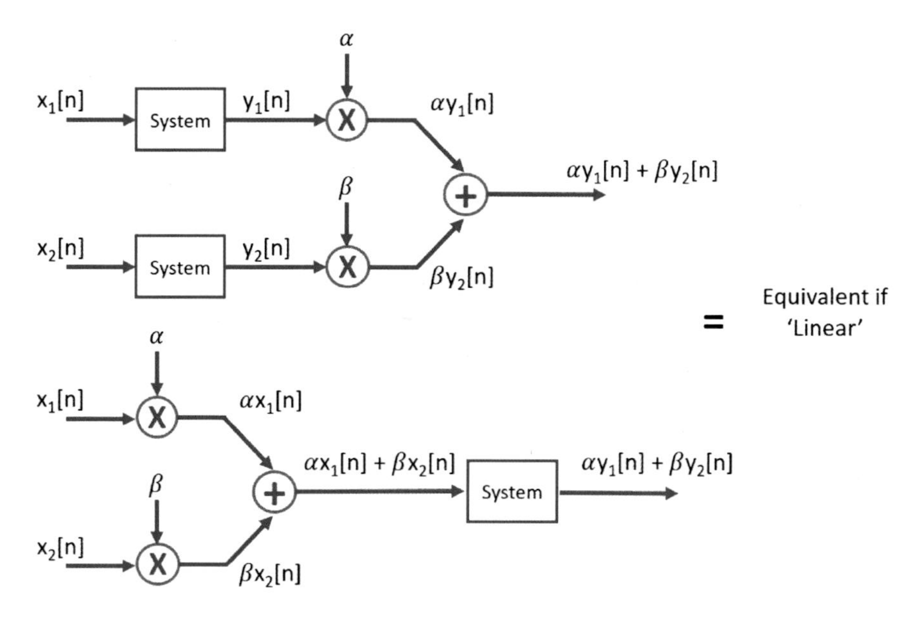

An interesting observation to be made here is that, for linear systems zero input yields zero output. Linearity of a system is determined using this property along with the additivity and the homogeneity property. If any of these conditions fail, then the system is non-linear.

2.4 LTI SYSTEMS

Both Linear and Time Invariant systems are two very important classes of systems, but Real world systems are seldom Linear and Time Invariant in nature. Most of the practical systems are non-linear to a certain extent. However, more often than not, we model real world systems as Linear Time Invariant systems or LTI systems. This is because, analyzing and finding solutions to non-linear time variant systems are extremely difficult and time consuming. But if the presence of certain non-linearity is negligible and not effecting the system response badly, they can be treated as Linear (for limited range of operation) by making some assumptions and approximations. This makes the math a lot easier and allows us to use more mathematical tools for analysis or in Richard Feynman's words "*Linear systems are important because we can solve them*". The advantages of making this approximation is far greater than any disadvantages that arises from the assumption. Even highly non-linear systems are treated as LTI for analysis and the non-linearity adjustments are made later. Any system that we refer from this point on in this book will be an LTI system. Several Properties of the LTI system, including the all-important Convolution property is discussed in the upcoming chapters.

3. FOURIER ANALYSIS

The name is Fourier, Joseph Fourier! In 1807, French physicist and mathematician Joseph Fourier (pronounced Fouye) came up with a crazy idea that gave a whole new meaning to signal processing. The idea was so crazy that, even other famous mathematicians of the time, like Lagrange opposed it. Fourier analysis is the back bone of the DSP and there's no way around it. Fourier analysis is math intensive, but in this chapter we have tried to make it as intuitive as possible and only the at most important topics are discussed.

Jean Baptiste Joseph Fourier

3.1 FOURIER SERIES

Fourier series is a basic mathematical tool for representing periodic signals. Using Fourier series it is possible to express periodic signals as some combination of sinusoids. This is applicable to any periodic function, however awkwardly shaped they may be. Isn't this cool?

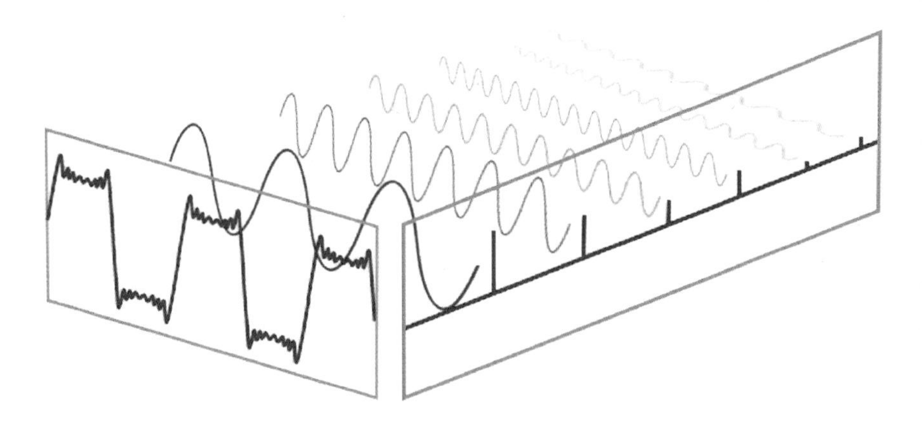

The above figure perfectly explains the Fourier series. Notice how a series of sinusoids combine to form the resultant signal, which looks nothing like a sinusoid. These components have different amplitudes and different frequencies.

Let's make things more interesting. Remember the superposition property of LTI systems from the previous chapter, this is where it comes in handy. The superposition property states that the response of a linear system to a sum of signals is the sum of the responses to each individual input signal. So instead of using a single signal as the input to a system, why not input component sinusoids to the system and add up their responses. Wouldn't both be the same? Yes, it would be, thanks to Fourier series. So the only thing we really need to know is the response of the system to sinusoids. From this we can predict the response to any other periodic signals.

$$x(t) = a_0 + \sum_{n=1}^{\infty} [a_n \cos(n\omega t) + b_n \sin(n\omega t)]$$

here $\mathbf{a_0,a_1,a_2...,b_1,b_2,b_3}$....are the Fourier coefficients. They tell us how much a sine or cosine signal of a particular frequency is contributing to the resultant signal.

The value of $\mathbf{a_0}$ tells us how much a cosine of zero frequency (**cos 0 =1**) or constant term is present in the resultant signal. $\mathbf{a_0}$ is also called the DC value or the Average value or the DC offset. Since all the other terms in the expansion are pure sinusoids, they individually average to zero, so the average value of the whole signal solely depends on $\mathbf{a_0}$.

Since **sin 0 = 0**, there can't be any contribution from zero frequency sine function, so $\mathbf{b_0}$ is always **0**. The value of $\mathbf{a_1}$ tells us how much a cosine of fundamental frequency is present in the resultant function. Similarly, contribution from each sinusoid in the resultant function can be found out separately. This information is very useful, it can be used to manipulate function in a lot of ways.

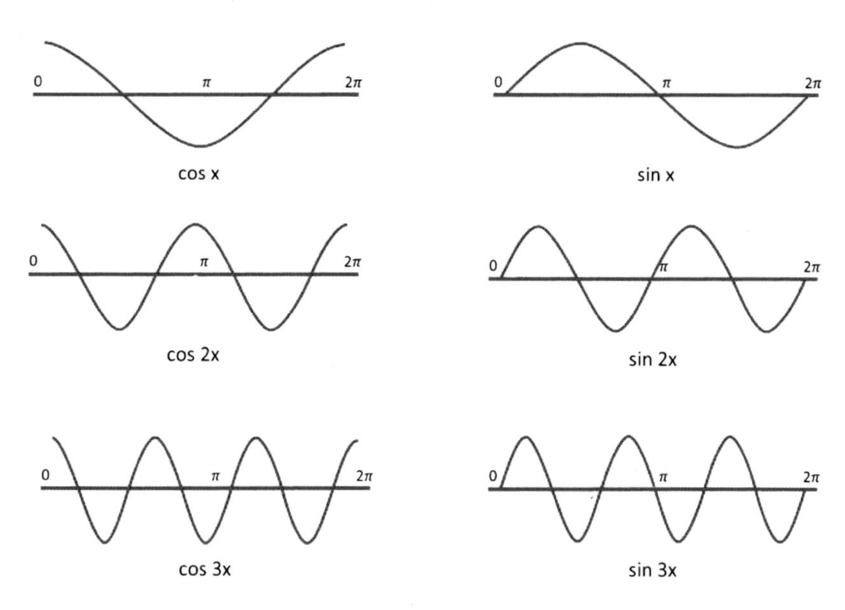

3.2 COMPLEX NOTATION

Fourier series can be expressed in a more compact form using complex notation. The contributions from both sine and cosine waves of the same frequency by a single coefficient.

$$x(t) = \sum_{n=-\infty}^{\infty} c_n \, e^{jn\omega t}$$

This is called the synthesis equation. Here the Fourier coefficients are complex. This notation has its own advantages, it is possible to calculate all Fourier coefficients using a single expression. Electrical engineers use **j** instead of **i**, since **i** is frequently used to denote electric current.

The values of $\mathbf{c_n}$ can be obtained using the expression:

$$c_n = \frac{1}{T} \int_0^T x(t) \, e^{-jn\omega t} \, dt$$

This expression is called the analysis equation and the plot of $|\mathbf{c_n}|$ vs **n** is called the frequency spectrum of the signal.

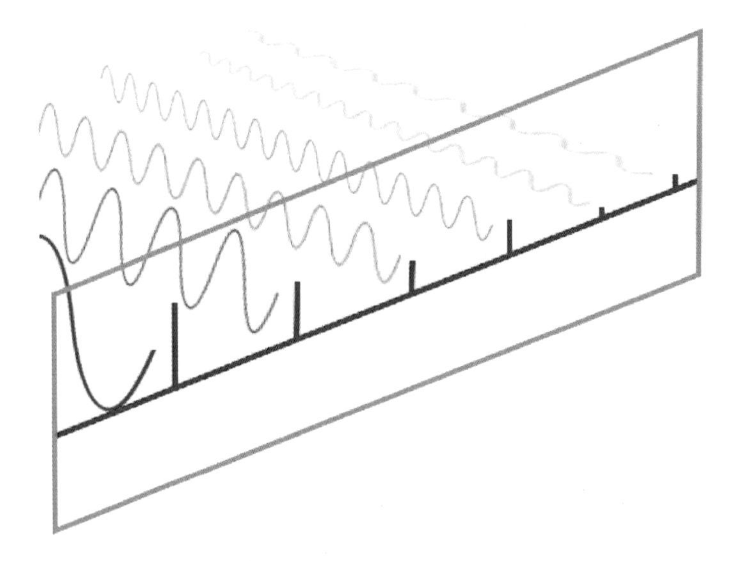

Notice the lines corresponding to each frequency component in the above picture. That is the Frequency spectrum. It tells us how much each frequency component contributes to the original signal. This information is invaluable to us.

Let's look at a practical example: In earth quake prone areas, houses are built to resist shock waves. But the earthquake is not a single frequency signal, it has many frequency components and it's not possible to design houses that are resistant to the entire wave. To overcome this difficulty, seismologists and structural engineers, do Fourier analysis on earthquake wave and use the frequency spectrum obtained, to figure out the dominant components in the wave. This way it is possible to design houses that are resistant to these particular frequency components in the wave. The other smaller components don't have any significant impact so they can ignored.

3.3 GIBBS PHENOMENON

The main reason behind Lagrange's objection to the Fourier series was that, he believed it is not possible to represent discontinuous functions (like square waves) in terms of sinusoids. Turns out there was some merit behind Lagrange's argument. In some way he was spot on, it is actually impossible to perfectly represent discontinuous signals using sinusoids.

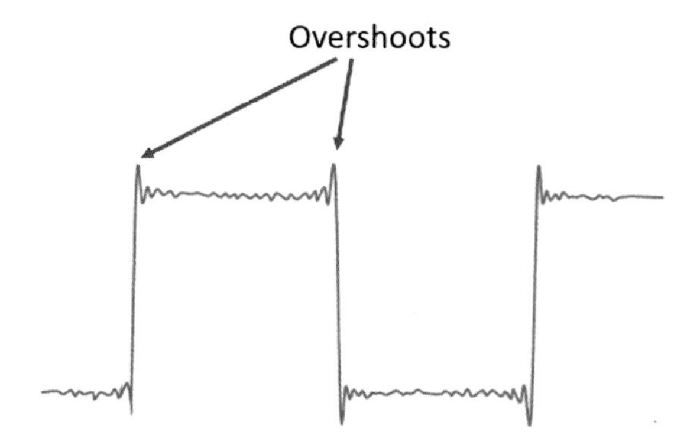

Notice how there is an overshoot at the corners of the square wave in the figure above. When a function takes a sudden jump, the Fourier estimation ends up overshooting that jump. This is known as Gibbs phenomenon. That overshoot will never go to zero no matter how many terms are added.

3.4 FOURIER TRANSFORM

We have now seen how the Fourier series is used to represent a periodic function by a discrete sum of complex exponentials. But how often are natural signals periodic?? Now that's a problem. Too bad we can't apply Fourier series to non-periodic signals. The way around this is to assume an aperiodic signal to be a periodic signal with infinite time period i.e. we are assuming the same pattern exists after infinite time.

This is where the Fourier transform comes in. The Fourier transform is used to represent a general, non-periodic function by a continuous superposition of complex exponentials. The Fourier transform can be viewed as the limit of the Fourier series of a function when the period approaches to infinity, so the limits of integration change from one period to $(-\infty, \infty)$.

The expression for the Fourier Transform is given by:

$$X(\omega) = \int_{t=-\infty}^{\infty} x(t)\, e^{-j\omega t}\, dt$$

For Periodic functions, there is a definite fixed no. of frequency components, but for Aperiodic functions there are infinite no. of frequency components, hence $\mathbf{X(\omega)}$ is a continuous function of ω, unlike $\mathbf{c_n}$ which is a discrete quantity. If you try using the Fourier transform on periodic functions, you can see that the Fourier series coefficients ($\mathbf{c_n}$) are basically the sampled values of $\mathbf{X(\omega)}$. In other words, $\mathbf{X(\omega)}$ forms the envelope for the Fourier series coefficients.

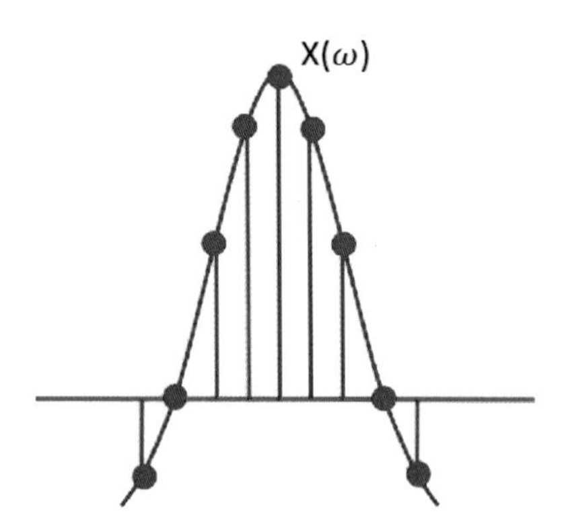

To obtain the original function from its frequency distribution, i.e. go back from frequency domain to the time domain, we need to use the Inverse Fourier transform:

$$x(t) = \frac{1}{2\pi} \int_{\omega=-\infty}^{\infty} X(\omega)\, e^{j\omega t}\, d\omega$$

3.5 PROPERTIES OF FOURIER TRANSFORM

3.5.1 Linearity

$$\text{if } x_1(t) \leftrightarrow X_1(\omega) \;\&\; x_2(t) \leftrightarrow X_2(\omega), \text{ then}$$

$$\alpha\, x_1(t) + \beta\, x_1(t) \leftrightarrow \alpha\, X_1(\omega) + \beta\, X_2(\omega)$$

3.5.2 Time shifting

If we were to time shift a signal, its magnitude spectrum won't change, only its phase spectrum changes.

$$\text{if } x(t) \leftrightarrow X(\omega), \text{ then}$$

$$x(t-t_0) \leftrightarrow X(\omega)\, e^{-jn\omega t_0}$$

Do note that the magnitude of $e^{-jn\omega t0}$= 1, so this term can only bring a phase shift.

3.5.3 Differentiation

$$\text{if } x(t) \leftrightarrow X(\omega), \text{ then}$$

$$x'(t) \leftrightarrow j\omega\, X(\omega)$$

3.5.4 Scaling Property

$$\text{if } x(t) \leftrightarrow X(\omega), \text{ then}$$

$$x(at) \leftrightarrow \frac{X(\frac{\omega}{a})}{|a|}$$

3.5.5 Duality

$$\text{if } x(t) \leftrightarrow X(\omega), \text{ then}$$

$$F\{X(\omega)\} \leftrightarrow x(-t)$$

For example, the Fourier transform of a square wave is a sinc function and the Fourier transform of a sinc function is a square wave.

3.6 MATLAB

3.6.1 Fourier series of a square wave with N harmonics

```
t = linspace(-2,2,10000);
f = 0*t;
N=7;

for n=-N:1:N
        if(n==0)                    % skip the zeroth term
                continue;
        end;

        C_n = ((1)/(pi*1i*n))*(1-exp(-pi*1i*n));            % computes the
n-th Fourier coefficient

        f_n = C_n*exp(2*pi*1i*n*t);             % n-th term of the series
        f = f + f_n;                            % adds the n-th term to f
end

plot(t, f, 'LineWidth', 2);
grid on;
xlabel('t');
ylabel('f(t)');
title(strcat('Fourier synthesis of the square wave function with n=',
int2str(N), ' harmonics.' ));
```

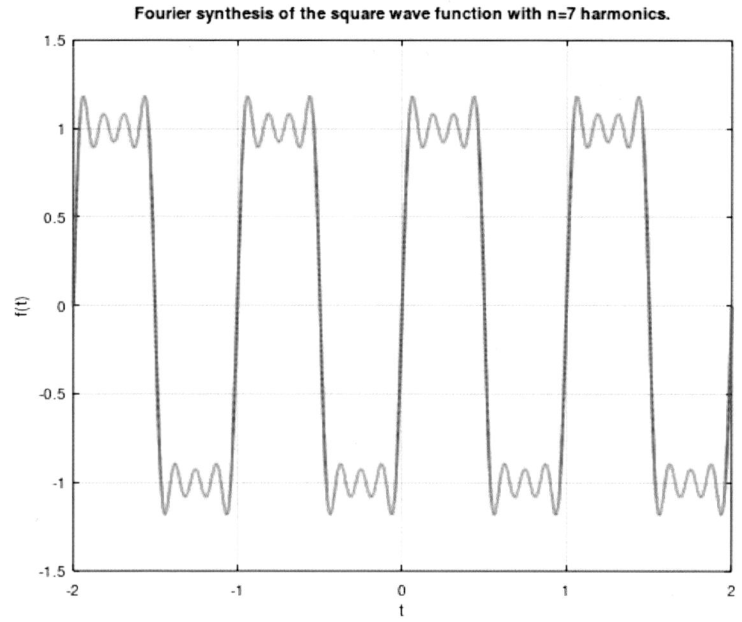

Try the same plot with a higher value for N (say 1000) and observe the Gibbs phenomenon in action.

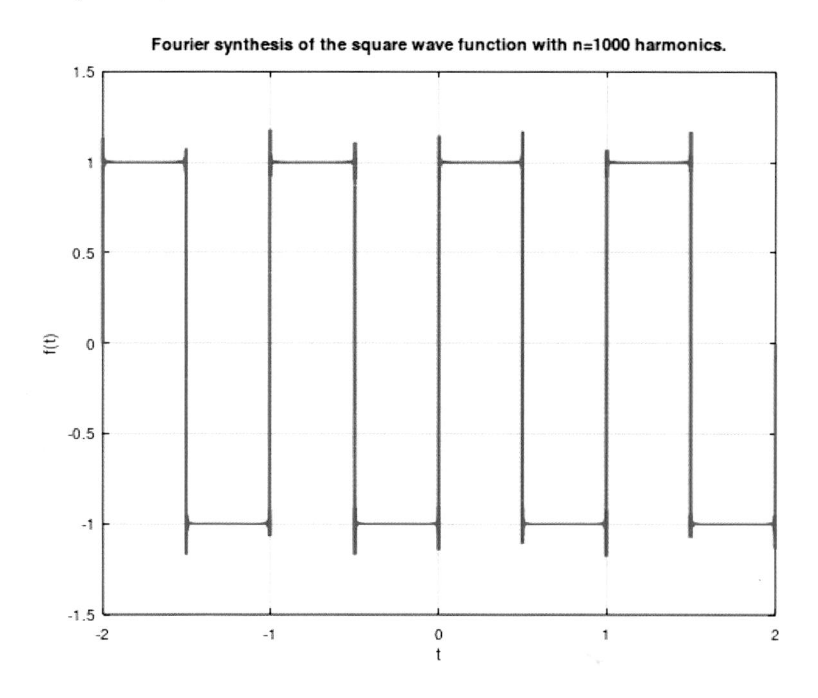

4. CONVOLUTION

4.1 SIFTING PROPERTY and THE IMPULSE RESPONSE

Convolution, the dreaded C word in DSP. Convolution may be the single most important concept in Signal Processing. So what really is Convolution? It's just a mathematical operation like addition or multiplication or division. The only difference is that these operators operate on numbers, whereas the convolution operator operate on signals.

How do we find out the response of a system to a signal or a range of signals? Surely we can't go testing the response of every signal one after the other. It's too cumbersome and many times impractical. There's got to be an easier way. That's where Convolution comes in. Convolution gives us the ability to predict the response of a system to a signal from a sample test result, or in other words, if we know the response of a system to any one input, then we can predict the system's response to any input using Convolution. This saves us some considerable effort since we are not actually doing the experiment, but at the same time we will have to deal with more math (which softwares like matlab handle with ease).

The benefits of Convolution will become more apparent as we go along. In any case we need to do the experiment at least once to obtain our sample test result to proceed with the Convolution. So the question is, what test signal do we use to obtain the sample result? Remember the good old Unit Impulse signal, it would be the perfect test signal. Bear in mind that there is no restriction on the test signal we use for convolution. But using the Unit Impulse signal has its advantages.

Consider this simple discrete triangular signal.

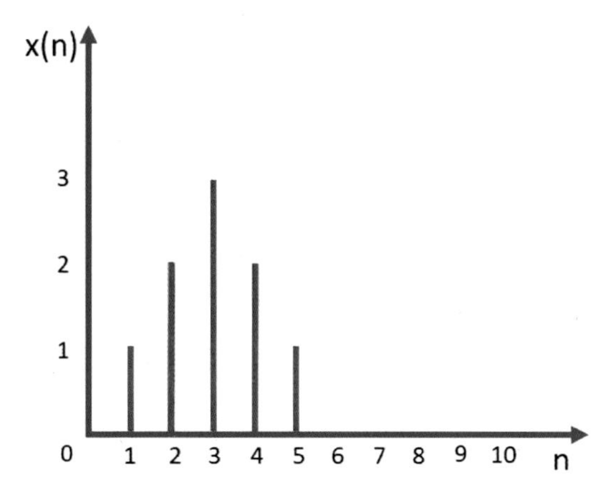

x[0] = 0, x[1] = 1, x[2] = 2, x[3] = 3, x[4] = 2, x[5] = 1

Now consider the 6 signals shown below.

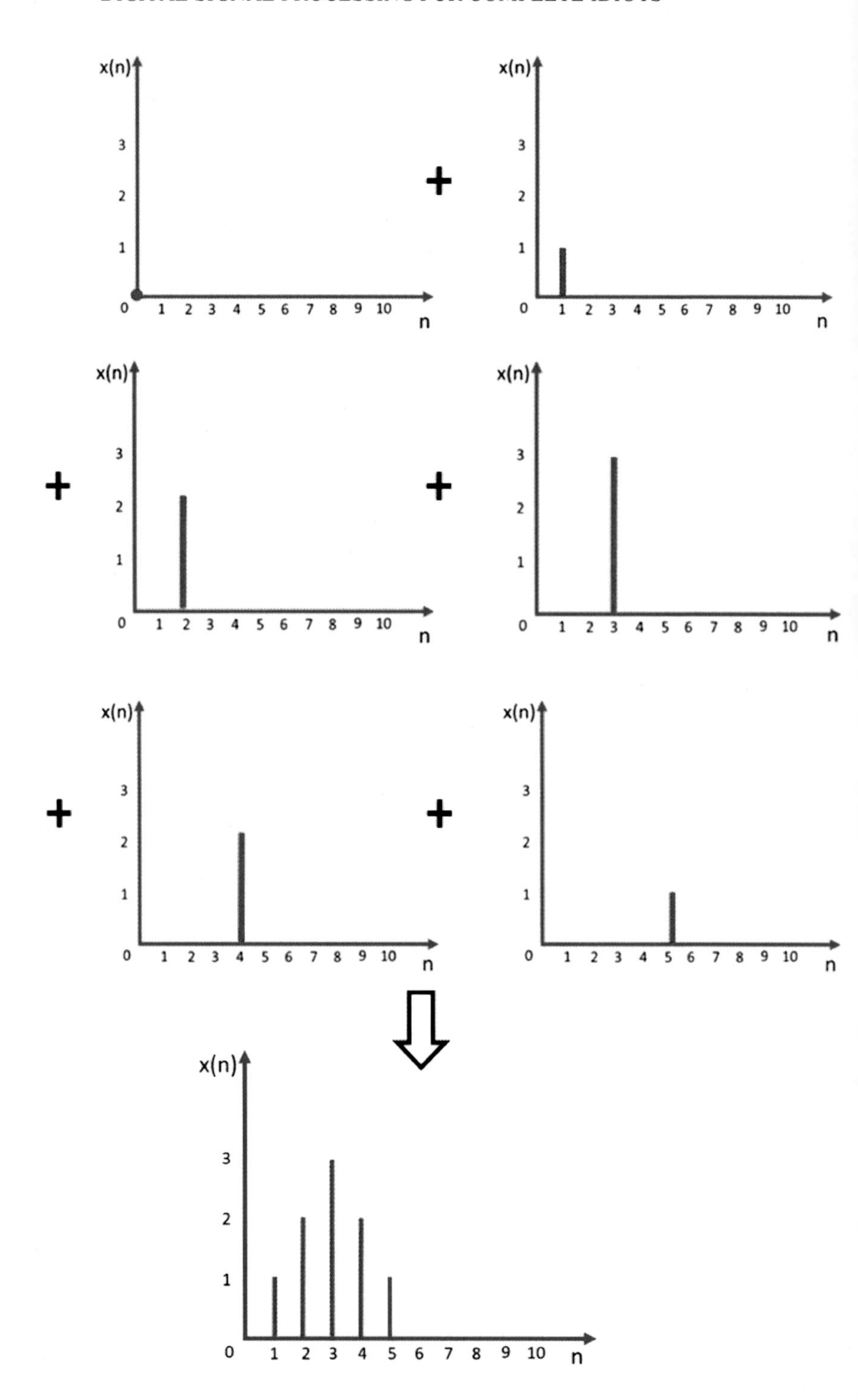
x(n)
3
2
1
0
1 2 3 4 5 6 7 8 9 10
n
+

These 6 signals are basically Unit Impulse signals that have been scaled and shifted in position and the discrete triangular signal from earlier is simply the sum of these 6 impulse signals. Mathematically, the triangular signal can be expressed as:

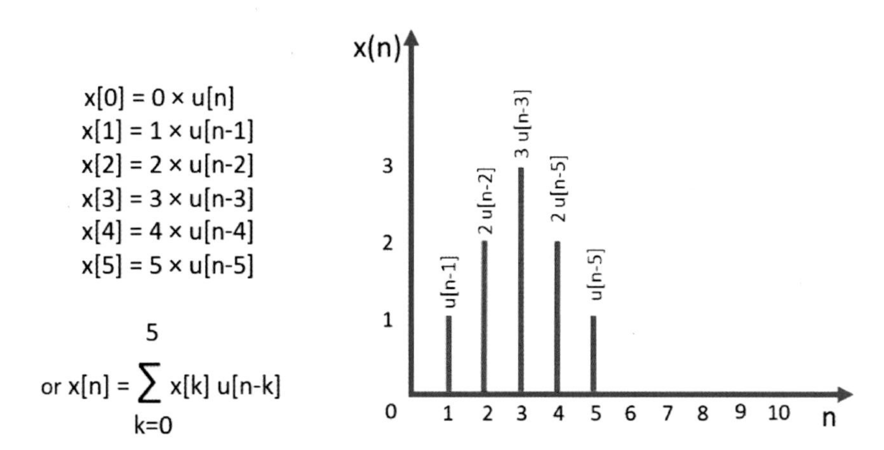

In this manner, any signal can be constructed out of scaled and shifted Unit Impulse signals. This is called the Sifting property of Unit Impulse function.

This is the main reason we use Unit Impulse signal as the test signal in convolution. By figuring out what a system does to a Unit Impulse signal, we can predict what the system does to any input signal. The response of a system to a Unit Impulse signal is called the Unit Impulse Response. It is denoted by **h[n]**.

Consider an example, let the Impulse Response of a system be as shown.

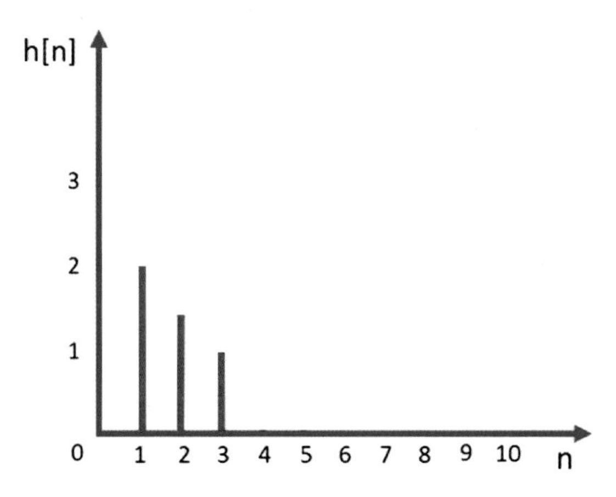

It is a common mistake to assume that the impulse response of a system is another Impulse, but that is mostly never the case. Imagine striking a bell with a hammer, the impact was just for a small time period (like our impulse function), but the ringing sound (the Impulse response) lasts for a while.

Now let's try to predict the response of the system to the input signal shown below.

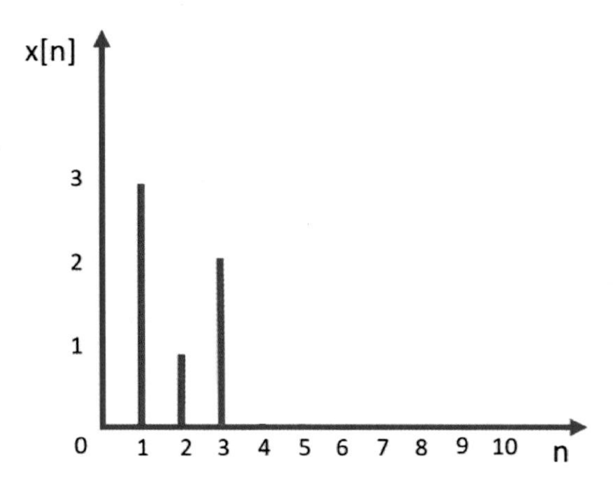

The input signal is basically made up of 4 Impulse signals of magnitudes 0, 3, 1, 2 respectively. The Impulse response is scaled by the same factor as the Impulse signal (this is a property of LTI systems). So let's look at the responses to these 4 impulses separately.

The first impulse is of zero magnitude, hence its response is also zero.

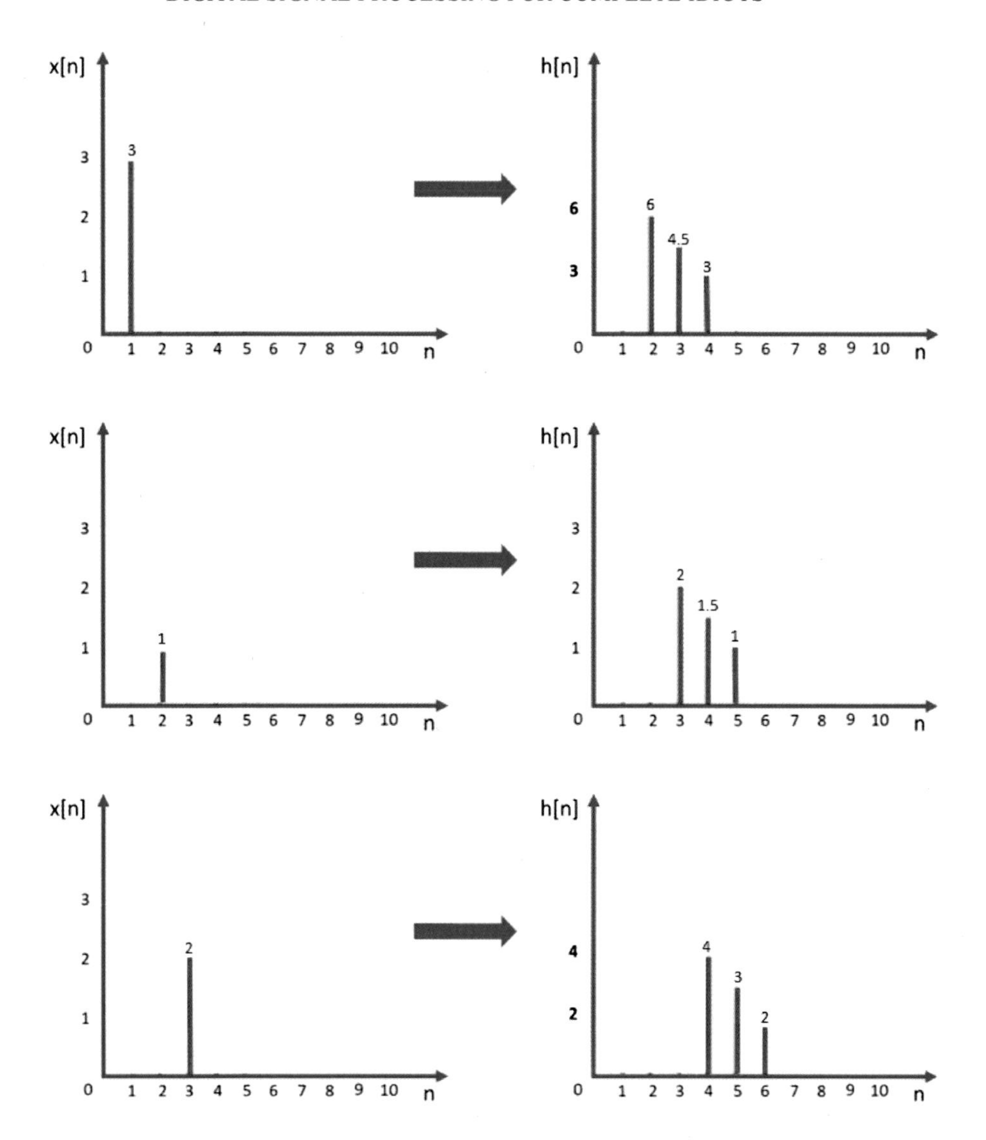

The response to the second impulse is the Unit impulse response scaled by 3 and shifted by 1. The response to the third impulse is the Unit impulse response scaled by 1 and shifted by 2. Finally, the response to the fourth impulse is the Unit impulse response scaled by 2 and shifted by 3. The output response is obtained by adding up these 4 impulse responses.

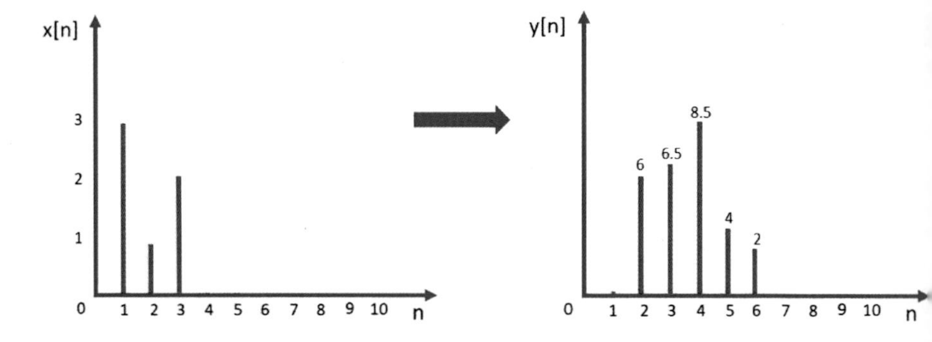

Voila !!! We have done it. We have predicted the output of a system to a signal from its Unit Impulse response.

This is the graphical way to perform convolution. In mathematical terms, convolution can be expressed as:

$$x[n] * h[n] = \sum_{k=-\infty}^{\infty} x[k]\, h[n-k]$$

Convolution operator is denoted by '*'.

For continuous signals, convolution is performed in a similar manner, as shown below.

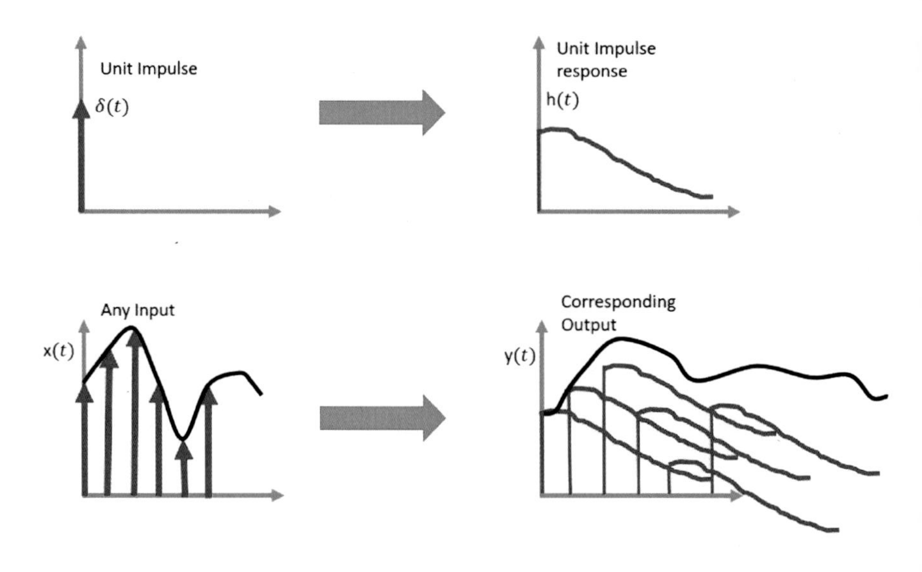

4.2 PROPERTIES OF CONVOLUTION

Convolution behaves in many ways like multiplication and has some similar properties.

4.2.1 Commutative

Convolution operator is commutative, meaning convoluting signal 1 with signal 2 is the same as convoluting signal 2 with signal 1.

$$x[n] * h[n] = h[n] * x[n]$$

Although it looks like the respective roles of x[n] and y[n] are different - one is "flipped and dragged" and other isn't - in fact they share equally in the end result. This property can be easily proved as follows:

$$x[n] * h[n] = \sum_{k=-\infty}^{\infty} x[k]\, h[n-k]$$

Now let n-k = l,

$$\therefore x[n] * h[n] = \sum_{l=-\infty}^{\infty} x[n-l]\, h[l]$$

$$= \sum_{l=-\infty}^{\infty} h[l]\, h[n-l] = h[n]*x[n]$$

4.2.2 Associative

Convolution operator is associative, which means that if convolution is performed among 3 or more signals, the order in which the convolution is performed is insignificant.

$$\{ x[n] * h_1[n] \} * h_2[n] = x[n] * \{ h_1[n] * h_2[n] \}$$

Associative property can be proved in pretty much the same way as we proved Commutative property.

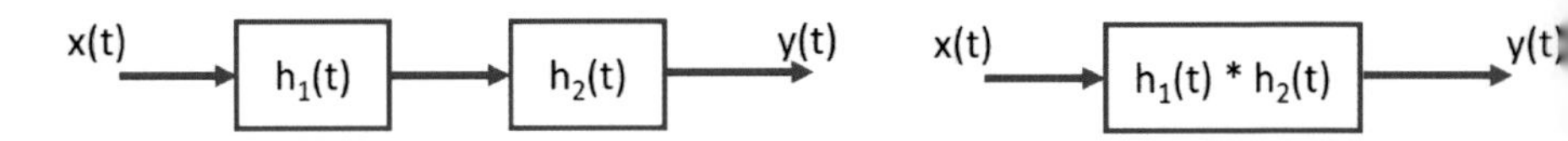

4.2.3 Distributive

Convoluting a signal with the sum of two signals is the same as convoluting the signal with the two signals individually and adding up their results.

$$x[n] * \{ h_1[n] + h_1[n] \} = \{ x[n] * h_1[n] \} + \{ x[n] * h_2[n] \}$$

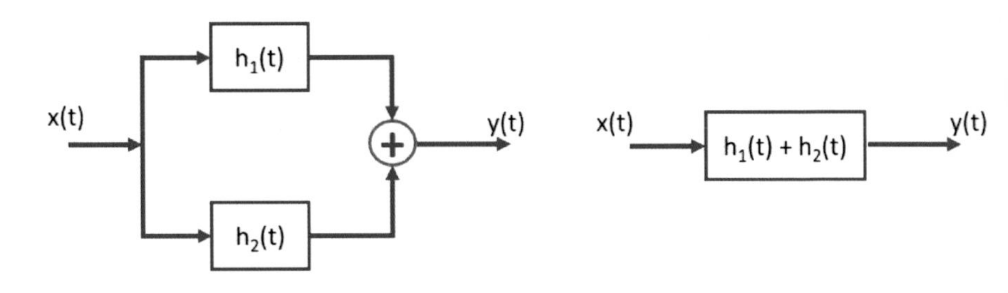

4.2.4 Convolution property

Convolution has a very interesting property, convolution in time domain corresponds to Multiplication in the frequency domain.

$$x[n] * y[n] = F(x[n]) \times F(y[n])$$

Where F denotes the Fourier transform.

We can combine these properties in whichever way we like and develop an algebra of convolutions. For example, the following property can be developed from the convolution property.

$$\{ x[n].y[n] \}*\{ f[n].g[n] \} = \{ F(x[n])*F(y[n]) \}.\{ F(f[n])* F(g[n]) \}$$

4.3 MATLAB

4.3.1 Convolution

```
x = input('Enter first sequence');
y = input('Enter second sequence');
z = conv(x,y);
subplot(3,1,1);
stem(x);
title('First input sequence');
subplot(3,1,2);
stem(y);
title('Second input sequence');
subplot(3,1,3);
stem(z);
title('Linear convolution');
```

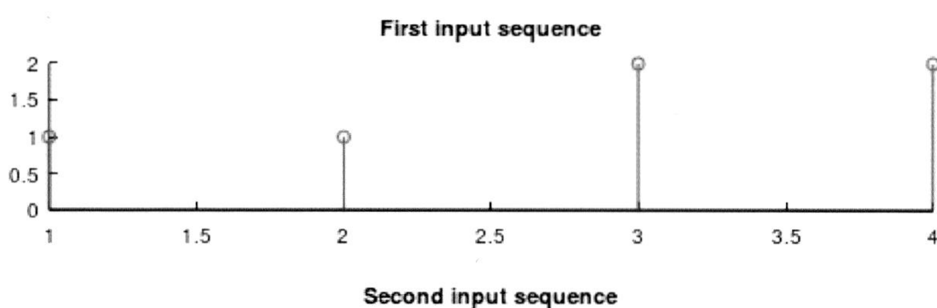

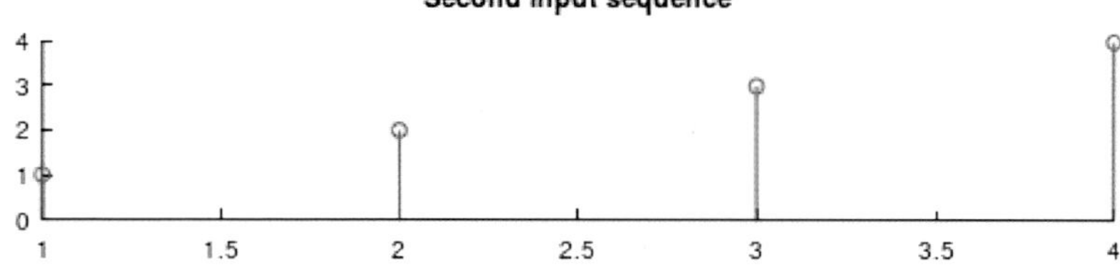

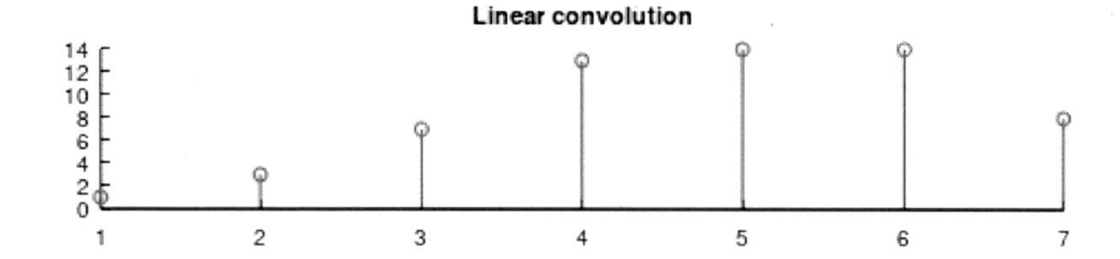

5. SAMPLING

5.1 CONTINUOUS TO DISCRETE

The signals that exist in real world are all analog or continuous time signals. Everything from sound to radiations from the sun are all analog in nature. But computers have a fixed memory and can only store a definite amount of data, so is not possible to process these signals directly using computers. In order to overcome this problem, we need to convert the analog signals to discrete form and the process of converting a signal from continuous time to discrete time is called Sampling. Sampling is done by measuring values of a continuous signal at certain intervals of time and neglecting the rest. Each measurement is called a Sample.

We do sampling because it is easier to process and manipulate a discrete time signal, but discrete signals are useful only for the in between processing stages. Once we are done processing a signal, at some point we would want to get back a continuous time (analog) output. For instance, when we edit a recorded audio, the whole editing process happens in the digital domain. But after editing, we need to use a speaker to get our output, which is in analog form. So the sampling process needs to be efficient to ensure reproducibility of the output continuous time signal after processing.

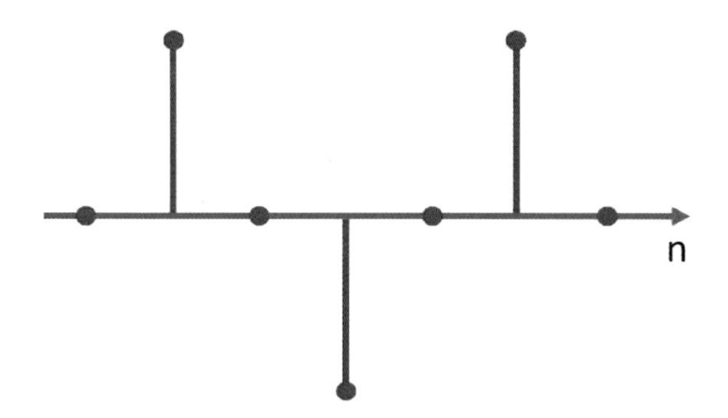

Consider the sample of a signal is shown in the figure above, let's try to reconstruct the original signal from this sample.

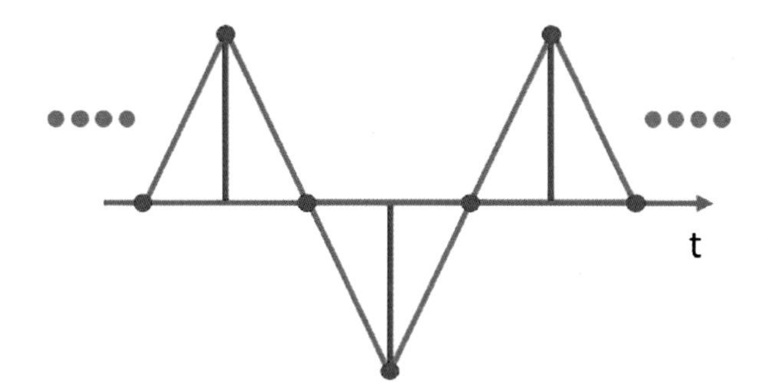

That was easy. But wait! Why not this signal shown below?

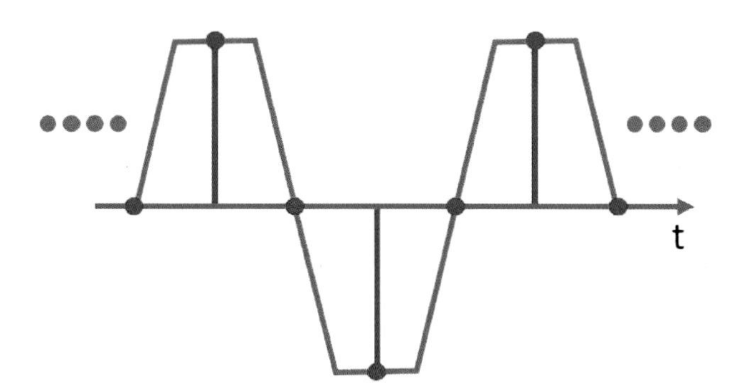

Or this one.

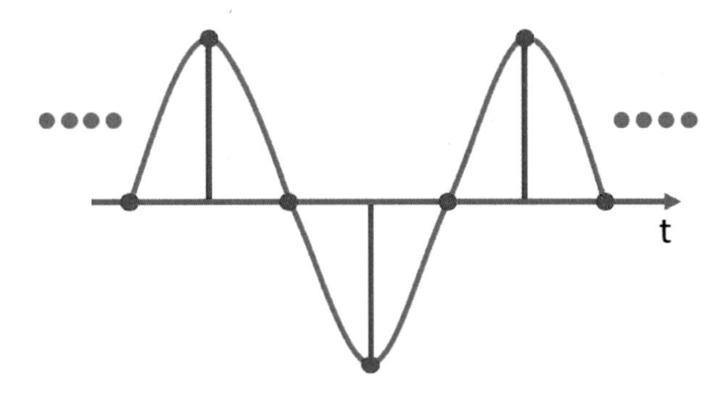

There seems to be endless possibilities. This is the problem with under Sampling. If the no. of samples are too low, it is very difficult or near impossible for faithful reproduction of the continuous time signal.

It's the sampling frequency that determines how faithfully we can reproduce an analog signal from its samples.

5.2 SAMPLING THEOREM

In the field of Digital signal processing, the Sampling theorem is the fundamental bridge between continuous-time signals and discrete-time signals. Engineers Harry Nyquist and Claude Shannon has been credited for the discovery of the Sampling Theorem, hence the name Nyquist-Shannon Sampling theorem.

Claude Shannon

Harry Nyquist

According to the Sampling theorem "**A Band-limited signal can be reconstructed without any error if it is sampled at a rate at least twice the maximum frequency component in it**" i.e. for perfect sampling, $f_s \leq 2f_m$. The minimum required sampling frequency $f_s=2f_m$ is called the Nyquist frequency.

A continuous time signal is sampled by multiplying it with an impulse train which is nothing but a series of impulses that periodically repeat.

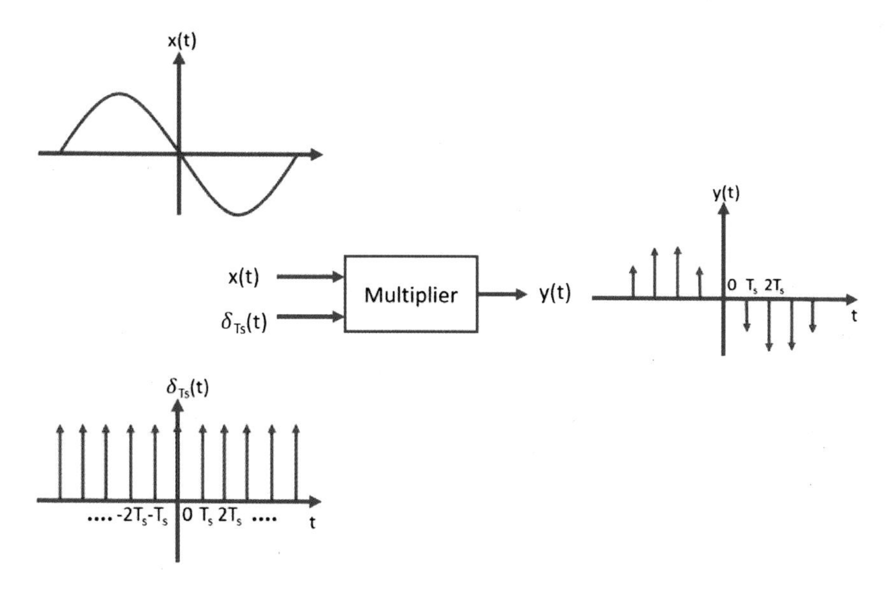

The Sampling Theorem can be better understood in the frequency domain. Firstly, let's define a "Band Limited" signal. A band-limited signal is any signal whose frequencies are limited within a particular range (band).

Suppose a band-limited signal $\mathbf{x(t)}$ whose bandwidth is $\mathbf{f_m}$ ($\omega_m = 2\pi f_m$) is to be sampled, then we need to multiply the signal by a sampling train $\boldsymbol{\delta_s(t)}$. The sampling train is mathematically defined as:

$$\delta_s(t) = \sum_{n=-\infty}^{\infty} \delta(t-nT_s)$$

So the sampled signal $\mathbf{X_s(t)}$ is given by,

$$X_s(t) = x(t) \times \delta_s(t)$$

$$= x(t) \times \sum_{n=-\infty}^{\infty} \delta(t-nT_s)$$

Taking Fourier transform on both sides, we can convert the equation to frequency domain (Remember multiplication becomes convolution in frequency domain).

$$X_s(\omega) = F\{ x(t) \times \delta_s(t)\}$$

$$= \frac{1}{2\pi} \times \{ x(\omega) * \delta_s(\omega)\}$$

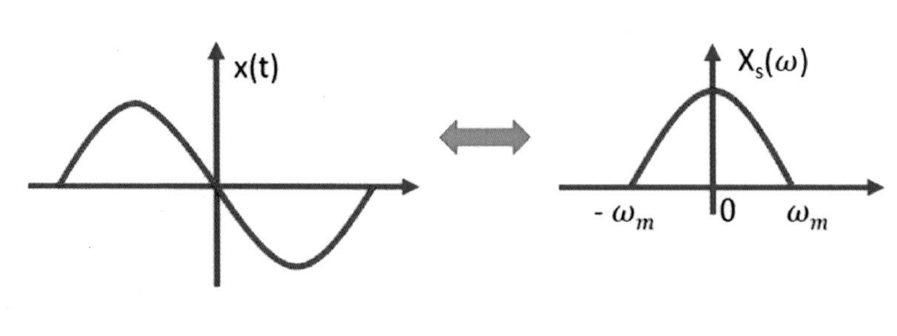

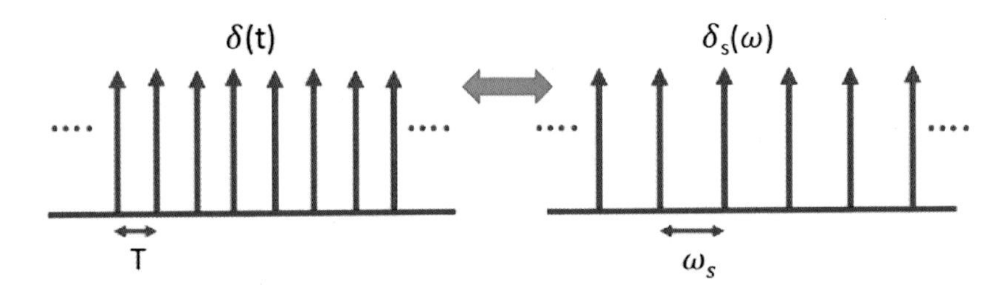

The impulse train in the frequency domain corresponds to another impulse train (this can be easily verified by taking the Fourier series)

$$\delta_s(\omega) = \frac{2\pi}{T_s} \sum_{n=-\infty}^{\infty} \delta(t - nT_s)$$

Therefore the sampled signal in the frequency domain is given by,

$$X_s(\omega) = \frac{1}{2\pi}\left\{ X(\omega) \ * \ \frac{2\pi}{T_s} \sum_{n=-\infty}^{\infty} \delta(t - nT_s) \right\}$$

Convolution of a signal with an impulse train will result in the signal taking the place of the impulses.

$$X_s(\omega) = \frac{1}{T_s} \sum_{n=-\infty}^{\infty} X(\omega - n\omega_s)$$

Now, say if you sample at a less than nyquist rate i.e. $f_s < 2f_m$, the high frequency components will interfere as shown below and lead to a phenomenon called Aliasing

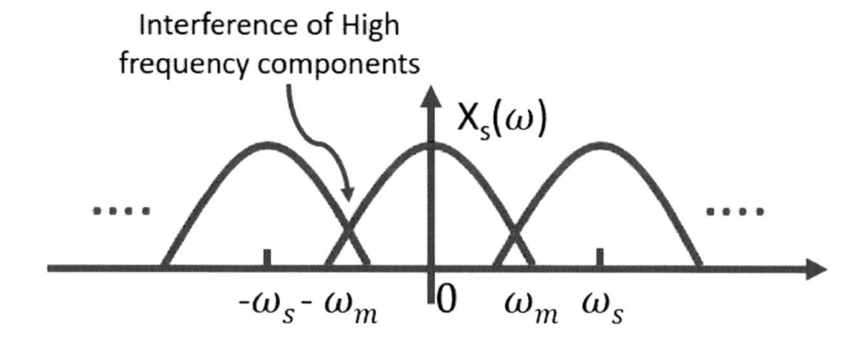

In theory, oversampling is a wasted effort, but in practice, a signal is oversampled or sampling is done above the Nyquist rate, just to be on the safe side.

6. DISCRETE FOURIER ANALYSIS

In the last chapter, we saw how a continuous time signal can be sampled to obtain its discretized version. Now we'll move on to processing such signals. In chapter 3, we used the Fourier series and the Fourier transform to obtain the frequency spectrum of continuous time signals. For processing discrete signals, we have to do the exact same thing with discrete signals, we need to analyze their frequency content. There are lots to tools to do that, like DTFT, DFT, z-transform etc. The tool of choice depends on the type of input signal and desired output.

6.1 DISCRETE TIME FOURIER TRANSFORM (DTFT)

DTFT or the Discrete time Fourier transform is used to analyze aperiodic discrete signals. The DTFT is analogous to the Fourier Transform used for Continuous time signals. It expresses a time domain signal in terms of complex exponential sequence $e^{j\omega n}$, where ω is a real frequency variable.

The Continuous time Fourier transform (CTFT) or simply Fourier transform is defined as:

$$X(\omega) = \int_{-\infty}^{\infty} x(t)\, e^{-j\omega t}\, dt$$

The DTFT has a similar form to CTFT. Since it is used for discrete signals, the integral will become summation as you would expect. The DTFT of a discrete signal **x[n]** is given by:

$$X(e^{j\omega n}) = X(\omega) = \sum_{n=-\infty}^{\infty} x[n]\, e^{-j\omega n}\, d\omega$$

X(ω) is a continuous complex function of ω and represents the frequency content of **x[n]**. **X(ω)** is a periodic function with period 2π, which implies that the frequency range of discrete time signal is unique over the frequency interval of $(0,2\pi)$ or $(-\pi, \pi)$. We can easily verify this property as follows:

$$X(\omega + 2\pi k) = \sum_{n=-\infty}^{\infty} x[n]\, e^{-j(\omega + 2\pi k)n}\, d\omega$$

$$= \sum_{n=-\infty}^{\infty} x[n]\, e^{-j\omega n}\, e^{-j2\pi kn}\, d\omega$$

As k and n are integers, $e^{-j2\pi kn} = 1$

$$\therefore X(\omega + 2\pi k) = \sum_{n=-\infty}^{\infty} x[n]\, e^{-j\omega n}\, d\omega$$

$$= X(\omega)$$

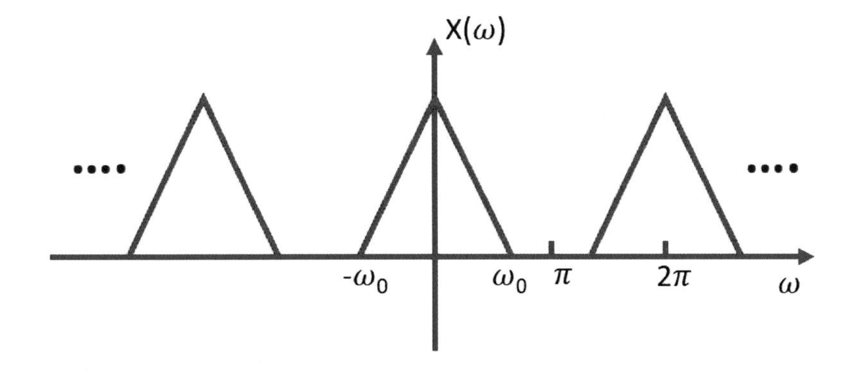

Like with any other complex function, **X(ω)** can be expressed in the form: **X(ω)** = **|X(ω)|∠ X(ω)**. The plot between **|X(ω)|** and **ω** is called the Magnitude spectrum and Plot between **∠ X(ω)** between **ω** is called the Phase spectrum.

6.2 INVERSE DTFT

To obtain the time domain signal from frequency spectrum, we need to use IDTFT or the Inverse Discrete Time Fourier Transform. In continuous time domain, the Inverse Fourier transform is given by:

$$x(t) = \frac{1}{2\pi} \int_{\omega=-\infty}^{\infty} X(\omega)\, e^{j\omega t}\, d\omega$$

By analogy to the Continuous time signals, the IDTFT can be written as:

$$x[n] = \frac{1}{2\pi} \int_{\omega=-\pi}^{\pi} X(\omega)\, e^{j\omega t}\, d\omega$$

When we wrote the DTFT formula by analogy, we changed the integral to summation. Why not here? Well, x[n] is a discrete function and **X(ω)** is a continuous function. So to obtain a continuous function from a discrete function we need to use summation and to obtain a discrete function from a continuous function we need to use an integral. Also note that we changed the interval to $(-\pi, \pi)$ in case of IDTFT, that's because **X(ω)** is periodic and the information outside this interval is redundant.

x[n] can be recovered uniquely from its DTFT, so they form a Fourier Pair: **x[n] ⇔ X (ω)**

6.3 DISCRETE FOURIER TRANSFORM

So far we have discussed the Fourier series and the Fourier transform for continuous time signals and DTFT for the Discrete time signals. Now we'll look into the Discrete Fourier Transform or DFT. The DFT is analogous to the Fourier series used for continuous time signals.

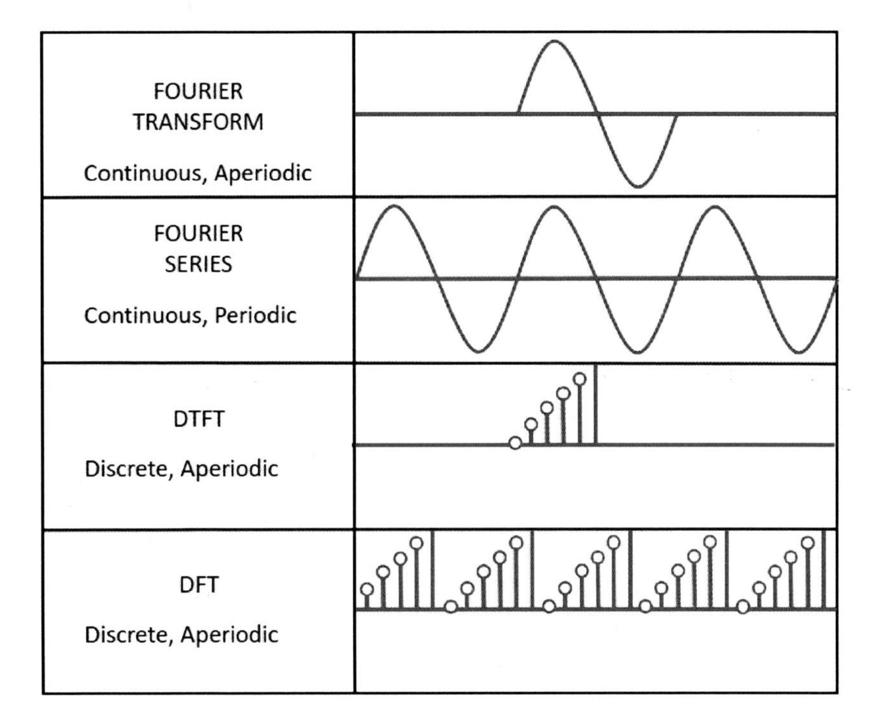

You must be thinking that the names given to these four types of Fourier transforms are confusing and poorly organized. You're right, the names have evolved rather haphazardly over 200 years.

All these classes of Fourier transforms extend from negative to positive infinity. But computers can only a store finite no. of samples, so how can we apply these transforms? So is there a version of the Fourier Transform that uses finite length signals? No, there isn't. Sine and cosine waves are defined as extending from negative infinity to positive infinity and you cannot use a group of infinitely long signals to synthesize something finite in length.

What do we do now? We have to make the finite length signals look infinite someway. There are two ways to go about this. One method is to imagine that the signal has an infinite number of samples on the left and right of the actual points. If all these imaginary samples have a value of zero, the signal looks discrete and aperiodic, and the DTFT applies. The other method is to imagine that the samples on the left and right as the duplication of the actual stored samples (say 1024 samples). In this case, the signal looks discrete and periodic and hence DFT can be used.

But when it comes to practical situation, DTFT is not an option, because an infinite number of sinusoids are required to synthesize a signal that is Aperiodic. So it is not possible to implement it using computer algorithm. So, we are left with no choice. The only type of Fourier transform that can be used in DSP is the DFT.

The expression for the Fourier series (chapter 3) for continuous time signals is

$$X[k] = \frac{1}{T} \int_{0}^{T} x(t)\, e^{-jk\omega t}\, dt$$

By analogy, we can write the expression for DFT as:

$$X[k] = \sum_{n=0}^{N-1} x[n]\, e^{-jk(2\pi/N)n} \quad , k = 0,1,2....N-1$$

To make the notation a little easier, in DFT the exponential part (excluding k and n) is denoted as W_N and is called the Twiddle factor i.e. $\mathbf{W_N = e^{-j(2\pi/N)}}$. So the DFT can be written in terms of the Twiddle factor as:

$$X[k] = \sum_{n=0}^{N-1} x[n]\, W_N^{kn}$$

An N-point DFT essentially transforms a sequence of N complex numbers **x[1],x[2],x[3]** etc. into another sequence of N complex numbers **X[1],X[2],X[3]** etc.
(or two sequences of N/2 +1 real numbers, representing magnitudes of sine and cosine components respectively)

Just as the Fourier series is the sampled version of the Fourier transform, the DFT is the sampled version of the DTFT.

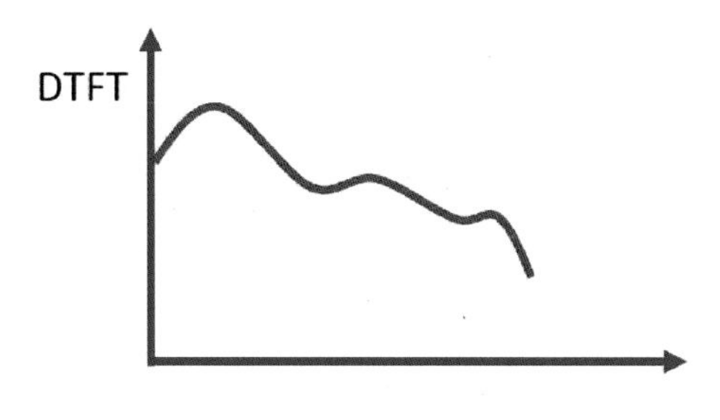

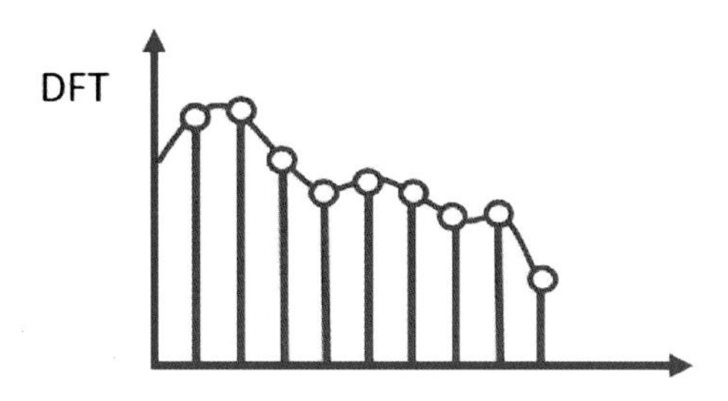

6.4 INVERSE DFT

The expression for the Inverse DFT is:

$$x[n] = 1/N \sum_{k=0}^{N-1} X[k]\, W_N^{-kn}$$

6.5 PROPERTIES OF DFT

Here are some of the basic properties of DFT:

6.5.1 Linearity

Linearity is a property common to every member in the Fourier Transform family.

$$\text{if } x_1[n] \leftrightarrow X_1[k] \text{ and } x_2[n] \leftrightarrow X_2[k],$$

$$\text{then, } ax_1[n] + bx_2[n] \leftrightarrow aX_1[k] + bX_2[k]$$

6.5.2 Periodicity

if x[n] and X[k] are an N-point DFT pair,

then, $x[n + N] = x[n]$ & $X[k + N] = X[k]$

This can be proved as follows:

$$x(n + N) = \frac{1}{N} \sum X[k]\, e^{jk(2\pi/N)n}$$

Therefore,

$$x(n + N) = \frac{1}{N} \sum X[k]\, e^{jk(2\pi/N)n} .\, e^{jN(2\pi/N)n}$$

$$= \frac{1}{N} \sum X[k]\, e^{jk(2\pi/N)n} .\, e^{j2\pi n}$$

We know that, $\mathbf{e^{j2\pi k} = \cos 2\pi + j \sin 2\pi = 1}$
Substituting this result,

$$x(n + N) = \frac{1}{N} \sum X[k]\, e^{jk(2\pi/N)n} = X[k]$$

Similarly, periodicity in frequency domain can also be proved. (Do remember, $\mathbf{e^{-j2\pi k}=1}$)

6.5.3 Time Shifting

If $x[n] \leftrightarrow X[k]$,

then, $x[n - n0] \leftrightarrow X[k]\, e^{jk(2\pi/N)\, n0}$

Properties like Duality, Parseval's Theorem etc. are applicable to DFTs also. Some more properties like the Circular Convolution are discussed in the next chapter.

6.6 MATLAB

6.6.1 DFT (without the inbuilt function)

```
x=input('Enter the sequence x= ');
N=input('Enter the length of the DFT N= ');
len=length(x);
if N>len
  x=[x zeros(1,N-len)];
elseif N<len
   x=x(1:N);
end
i=sqrt(-1);
w=exp(-i*2*pi/N);
n=0:(N-1);
k=0:(N-1);
nk=n'*k;
W=w.^nk;
X=x*W;
disp(X);
subplot(211);
stem(k,abs(X));
title('Magnitude Spectrum');
xlabel('Discrete frequency');
ylabel('Amplitude');
grid on;
subplot(212);
stem(k,angle(X));
title('Phase Spectrum');
xlabel('Discrete frequency');
ylabel('Phase Angle');
grid on;
```

Output:

Enter the sequence x= [1 20 11 9]
Enter the length of the DFT N= 20

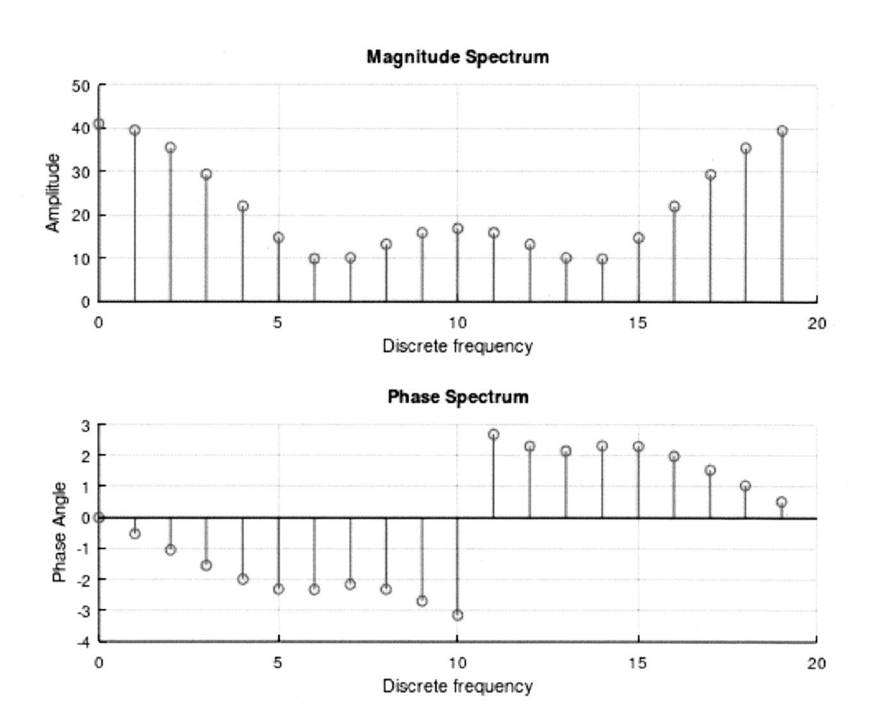
Magnitude Spectrum
Amplitude
50
40
30
20
10
0
0
5
10
15
20
Discrete frequency
Phase Spectrum
Phase Angle
3
2
1
0
-1
-2
-3
-4
0
5
10
15
20
Discrete frequency

7. FAST FOURIER TRANSFORM

7.1 NEED FOR SPEED

Let's revisit the DFT equation from the last chapter:

$$X[k] = \sum_{n=0}^{N-1} x[n]\,(W_N)^{nk}$$

When computing DFT, to obtain one value of X[k], we need to do N complex multiplications and (N-1) complex additions and to obtain N values of X[k], we need to do N^2 complex multiplications and N(N-1) complex additions. As useful as DFT is, as N gets bigger, the process becomes inefficient and time consuming. So direct evaluation of the DFT is probably not a good idea. This is where FFT or the Fast fourier transform comes in. The FFT is not another type of Fourier Transform, it's rather an algorithm (a method) to make DFT's faster and more efficient. FFT algorithms were popularized by Cooley and Tukey.

James William Cooley

John Wilder Tukey

The main idea behind the FFT is to decompose the DFT to smaller DFT's. The FFT also makes use of some of the properties of the Twiddle factor (W_N).

There are basically two classes of FFT algorithms: Decimation in time (DIT-FFT) and Decimation in frequency (DIF-FFT). In DIT-FFT, the sequence for which we need the DFT is successively divided in smaller sequences and the DFT's of the subsequences are combined to obtain the required DFT. In DIF-FFT, the frequency samples of the DFT are decomposed into smaller subsequences.

7.2 DECIMATION IN TIME ALGORITHM

In FFT, we are expressing the number of output points N as a power of 2, i.e. $N = 2^M$. Bear in mind that the DFT is a sampled version of the DTFT, so we are free to choose the value of N. So say if N = 900, we can add 124 zeros and make $N = 1024 = 2^{10}$ .We could have also neglected 388 points and made $N= 512 =2^{9}$, but that would mean losing some information. So the first method is preferred.

$$X[k] = \sum_{n=0}^{N-1} x[n]\,(W_N)^{nk}$$

Separating the even and odd terms, we can write X[k] as,

$$X[k] = \sum_{n\ \text{even}}^{N-1} x[n]\,(W_N)^{nk} + \sum_{n\ \text{odd}}^{N-1} x[n]\,(W_N)^{nk}$$

Let **n = 2r**

$$X[k] = \sum_{r=0}^{N/2-1} x[2r]\,(W_N)^{2rk} + \sum_{r=0}^{N/2-1} x[2r+1]\,(W_N)^{(2r+1)k}$$

$$= \sum_{r=0}^{N/2-1} x[2r]\,(W_N^{2})^{rk} + (W_N)^{k} \sum_{r=0}^{N/2-1} x[2r+1]\,(W_N^{2})^{rk}$$

$\mathbf{W_N^2 = W_{N/2}^1}$ (since $\mathbf{e^{-j2\pi nk/N} = e^{-j\pi nk/(N/2)}}$)

$$X[k] = \sum_{r=0}^{N/2-1} x[2r]\,(W_{N/2})^{rk} + (W_N)^{k} \sum_{r=0}^{N/2-1} x[2r+1]\,(W_{N/2})^{rk}$$

Therefore, $\mathbf{X[k] = G[k] + W_N^k H[k]}$. This way we can express an N-point DFT as the sum of two N/2-point DFT's.

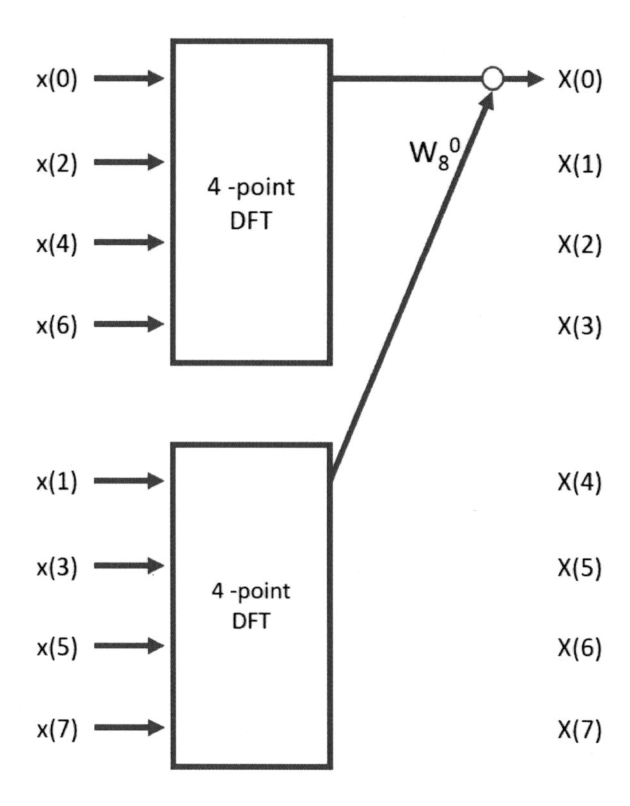

In the figure shown above, we have expressed an 8-point DFT as two 4-point DFT's. Going by the formula we derived, each term can be obtained as:

$\mathbf{X[0] = G[0] + W_8^0 H[0]}$ (shown in the figure)
$\mathbf{X[1] = G[1] + W_8^1 H[1]}$
$\mathbf{X[2] = G[2] + W_8^2 H[2]}$
$\mathbf{X[3] = G[3] + W_8^3 H[3]}$
$\mathbf{X[4] = G[0] + W_8^4 H[0]}$
$\mathbf{X[5] = G[1] + W_8^5 H[1]}$
$\mathbf{X[6] = G[2] + W_8^6 H[2]}$
$\mathbf{X[7] = G[4] + W_8^7 H[3]}$

Do note that **G[0] = G[4], G[1] = G[5], G[2] = G[6], G[3] = G[7]** and **H[0] = H[4], H[1] = H[5], H[2] = H[6], H[3] = H[7]**, since they are 4-point DFT's.

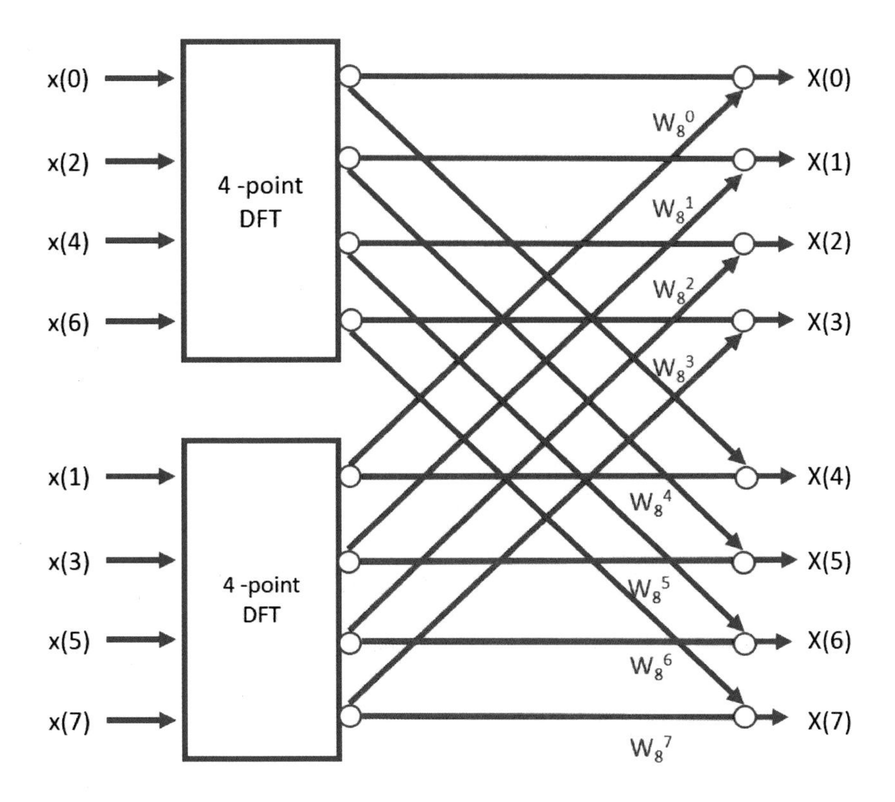

The question is, what is the advantage of doing things this way? We already know that, to compute an N-point DFT we need to do N^2 complex multiplications. In this case, we are computing two 4 -point DFT's, so we would need to do 32 complex multiplications, as opposed to the 64 complex multiplications if we had done a single 8-point DFT. That's a significantly improvement on the speed and efficiency of computation.

Now here's a thought, why don't we again split the 4 -point DFT's into two 2-point DFT's. That might save us some more time. Let's try that out.
We will separate the 4 -point DFT's into even and odd parts. Therefore, x[0] and x[4] form the even parts and x[2] and x[6] form the odd parts of the first 2-point DFT. Similarly, x[1] and x[5] form the even parts and x[3] and x[7] form the odd parts of the second 2-point DFT.

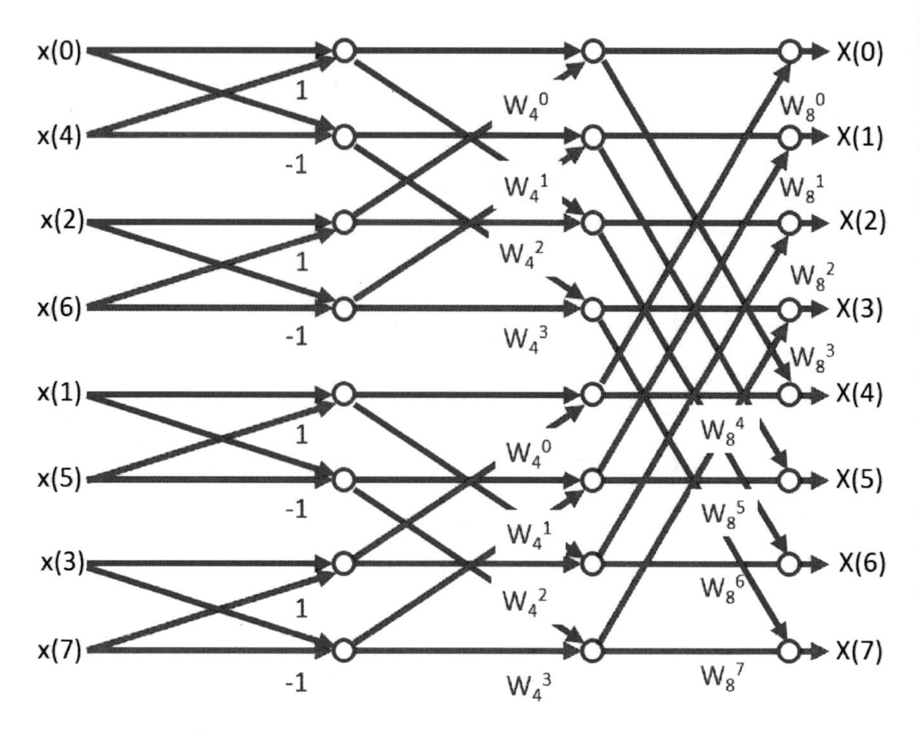

This way it is possible to make an N-point DFT into a number of 2 -point DFT's. The number of stages of operation required to obtain an N-point DFT is given by **$\log_2 N$**. For example, to obtain an 8-point DFT, it would take 3 stages of operation (as shown in the figure above). Using FFT algorithm we can bring down the number of complex multiplication to **$(N/2) \log_2 N$**. This may not seem much, but as N becomes larger, this is a significantly smaller number compared to **N^2**.

	Direct computation of DFT		FFT	
Number of Points	Complex Multiplications	Complex Additions	Complex Multiplications	Complex Additions
N	N^2	N^2-N	$(N/2)\log_2 N$	$N \log_2 N$
4	16	12	4	8
16	256	240	32	64
64	4096	4032	192	384
256	65536	65280	1024	2048
1024	1048576	1047552	5120	10240

A further simplification can be made to this algorithm, using the following property of the twiddle factor.

$$W_N^{r+N/2} = e^{-j(\frac{2\pi}{N})(r + N/2)} = e^{-j(\frac{2\pi}{N})r} \cdot e^{-j\pi}$$

$$= -W_N^r$$

Hence the butterfly unit can be modified as follows:

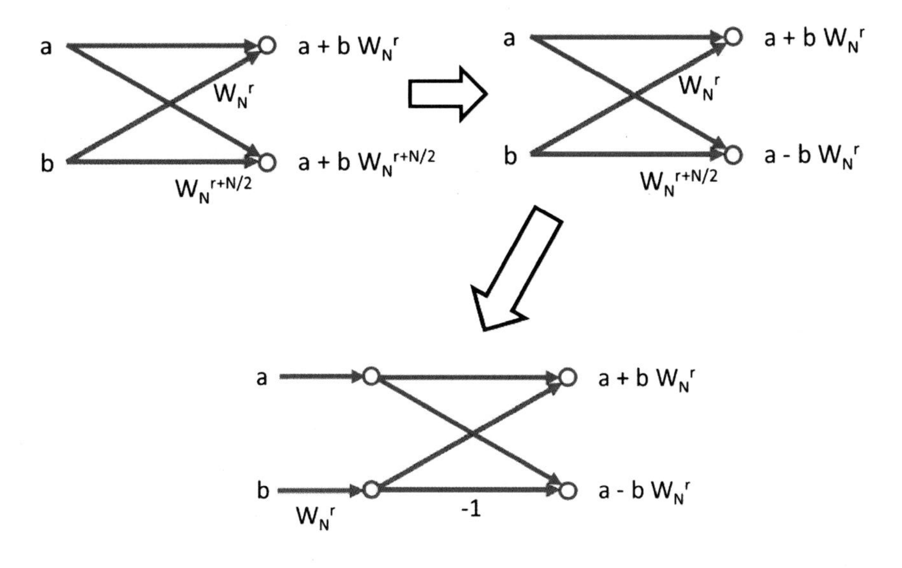

Using the simplified algorithm, the signal flow graph of an 8 point can be modified as shown below:

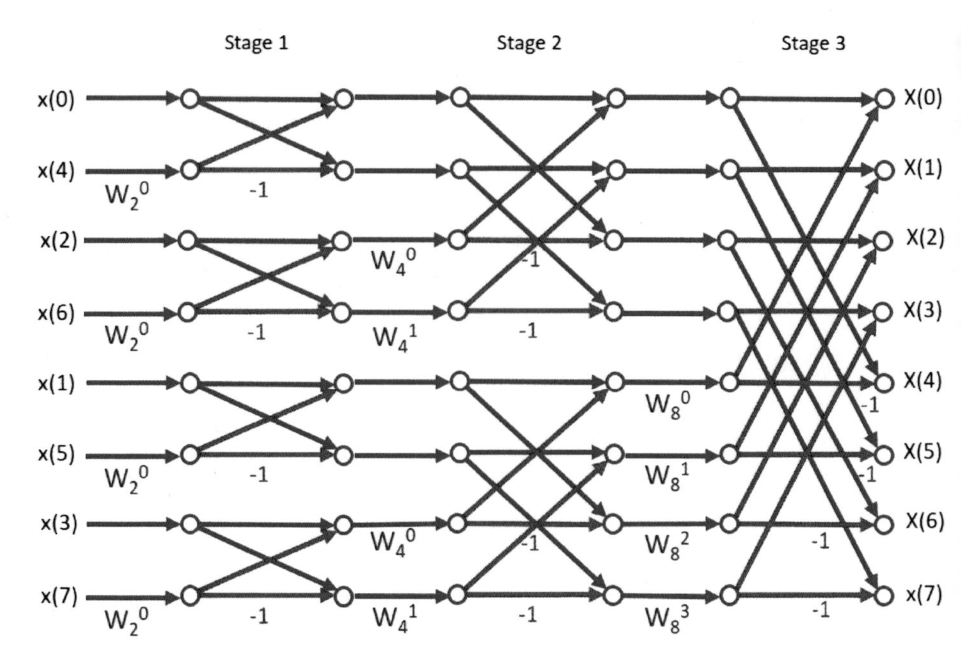

For the 8-point DFT, the inputs are arranged as follows: x(0), x(4), x(2), x(6), x(1), x(5), x(3), x(7). How do we figure out this order? For a higher order DFT, this becomes a hard task to memorize. Lucky for us there's an easy solution to this problem. Just write down the binary of these numbers in proper order and reverse their bits, and we get the desired order.

Normal order of index *n*	Binary bits of index *n*	Reversed bits of index *n*	Bit- reversed order of index *n*
0	000	000	0
1	001	100	4
2	010	010	2
3	011	110	6
4	100	001	1
5	101	101	5
6	110	011	3
7	111	111	7

7.3 DECIMATION IN FREQUENCY

In practice there is not much difference between DIT and DIF FFTs. In the DIF algorithm, the decimation is done in the frequency domain. That's why the frequency indices are in bit-reversed order. In this algorithm, you start with a single 8-point DFT, progress on to two 4-point DFTs and end with four 2-point DFTs.

Both the DIT-FFT and the DIF-FFT have the same complexity, but the hardware of some DSPs (Digital Signal Processors) are optimized for certain FFT variants.

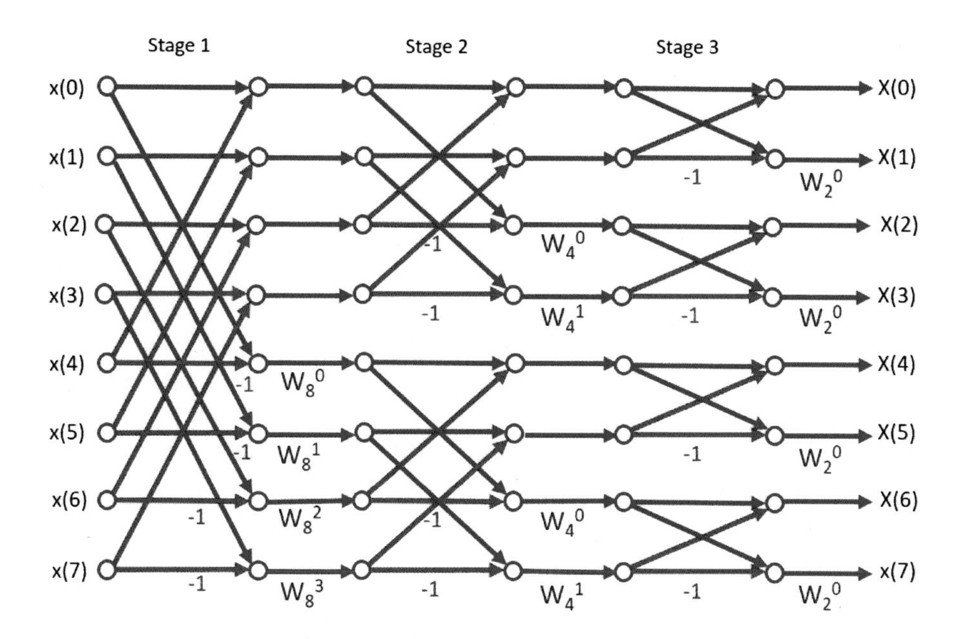

7.4 MATLAB

7.4.1 FFT

```
x = input('Enter the sequence');
n = input('Enter the length');
y= fft(x,n);
stem(y);
title('FFT');
xlabel('Real axis');
ylabel('Imaginary Axis');
```

Output:

Enter the sequence[1 20 11 9]
Enter the length20

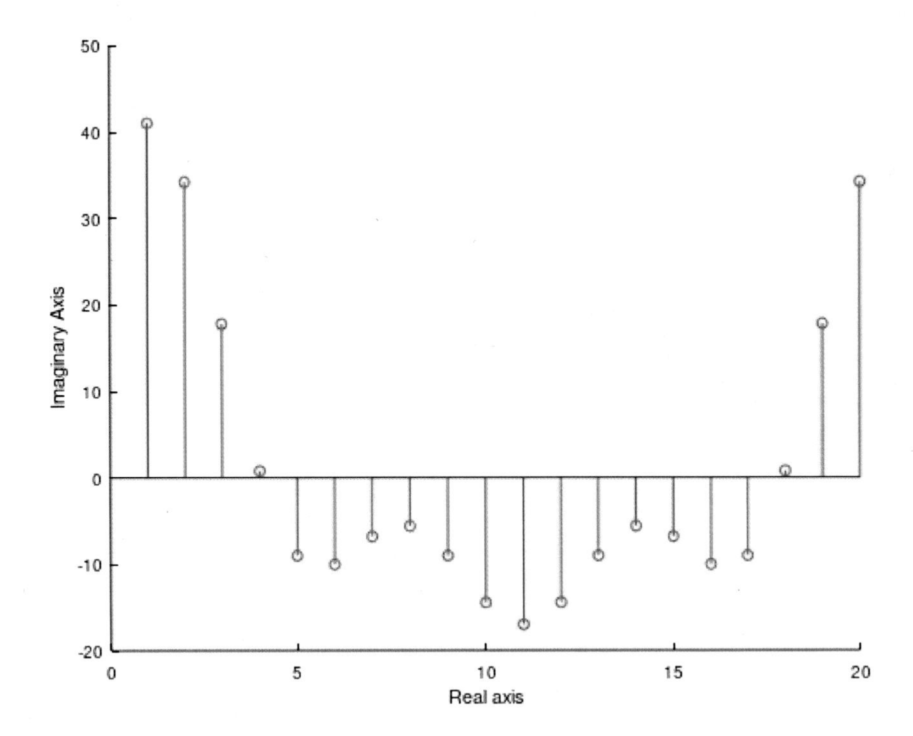

8. FREQUENCY RESPONSE

8.1 FREQUENCY RESPONSE

In one of the previous chapters, we saw how Convolution can be done in time domain. A similar analysis can be done in the frequency domain. But instead of analyzing the frequency response of a signal directly, the analysis is done on its component sinusoids, with the help of Fourier transform. So why do we need to use this indirect approach? Why not analyze the frequency response directly? There's a good reason for that, it's because of an interesting property of an LTI system, called sinusoidal fidelity. To explain sinusoidal fidelity, let's look at the mathematical operations associated with LTI systems. An LTI system can only be build using these or a combination of these operators:

- Multiplication by a constant: **f(t) x a**
- Differentiation of the input signal: **d f(t)/dt**
- Integration of the input signal: $\int$ **f(t) dt**
- Addition of two input signals: $\mathbf{f_1(t) + f_2(t)}$

Why are these operations so special? Let's try to find out, by providing sinusoidal inputs to these systems.

Let our sample sine input be: **A sin t** and the output of these operations will be:

Multiplication by a constant: **A sin t x a = Aa sint**. The output is a sine wave, just that the magnitude has just changed to **Aa**.

Differentiation of the input signal: **d(A sin t)/dt = A cos t = A sin(t +90)**. Again a sine wave output, this time the Magnitude doesn't change, but there is a phase shift.

This property is true for the other two operations as well. We have observed something very interesting here; when subject to sinusoidal input LTI systems produce sinusoidal output. So the shape of the input and the output waveforms are the same, the only things that change are the Amplitude or the Phase or both. This property is called Sinusoidal fidelity.

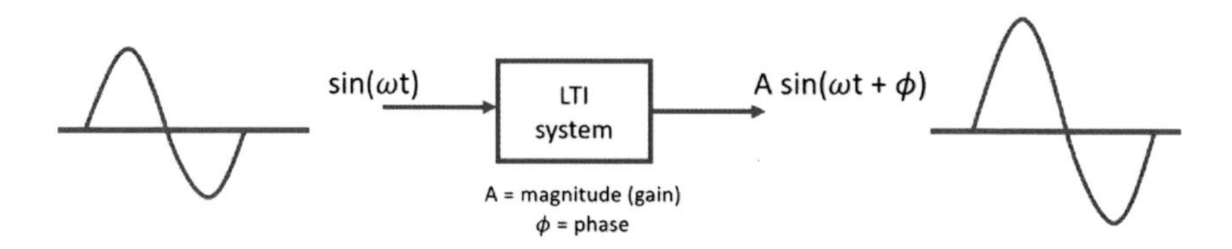

A critical inference can be made from this property, that both the input and the output signals will have the same frequency components i.e. No frequency components can be manufactured other than the one's already present in the input signal (ideally).

This means that any linear system can be completely described by how it changes the amplitude and phase of cosine waves passing through it. This information is called the system's Frequency response. Since both the Impulse response and the Frequency response contain complete information about the system, there is a one-to-one correspondence between the two. Given one, you can calculate the other. A system's frequency response is the Fourier Transform of its impulse response. H[n] is used to denote the frequency response.

8.2 CONVOLUTION via FREQUENCY DOMAIN

As mentioned earlier, Convolution in time domain corresponds to simple multiplication in frequency domain and vice versa i.e. **y[n] = x[n] * h[n]**, becomes **Y(ω) = X(ω) H(ω)** in frequency domain. Convolution is an extremely useful tool but there are some difficulties in using it. Firstly, the convolution operator is mathematically difficult to deal with. For instance, suppose you have a system's impulse response and its output signal, how do you calculate the input signal? The process is called Deconvolution, it is almost impossible to understand this operation in the time domain. However, Deconvolution can be carried out in the frequency domain as a simple division (inverse of multiplication). The second reason for avoiding convolution is computation speed. The standard convolution algorithm is slow because of the large number of multiplications and additions that must be calculated.

To tackle this problem, two signals that are to be convoluted are converted into the frequency domain, multiplied with each other, and then the result is transformed back into the time domain. This replaces one convolution operation with two DFTs, a multiplication, and an Inverse DFT. As far as speed is concerned, this approach isn't any faster. But that changed with the invention of the FFT algorithm.

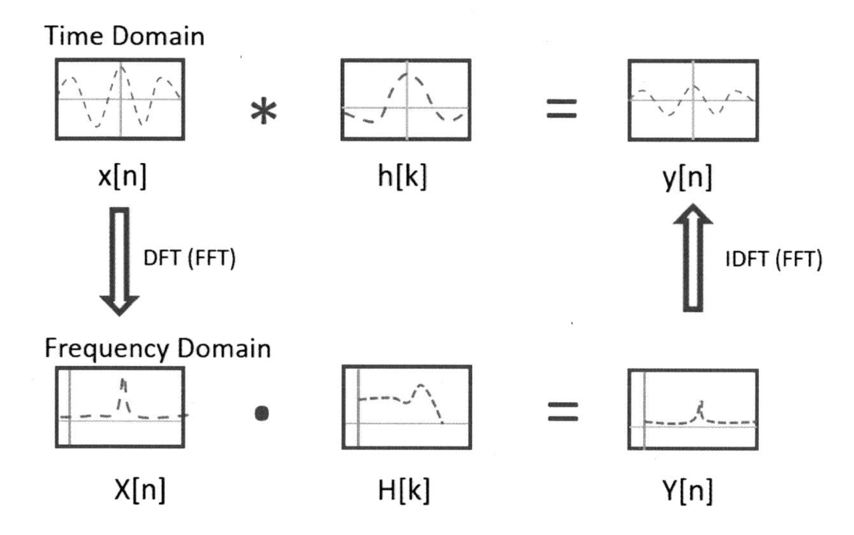

The Convolution of an N point signal with an M point impulse response results in an N+M-1 point output signal. In the example shown below, a 256 sample signal is convolved with a 51 sample impulse response, resulting in a 306 sample signal. If we use a 256-point DFT for convolution, the samples from 256 to 305 are pushed into the next period of the signal, distorting the signal. This process is called Circular convolution and should be avoided at all costs.

$$\mathrm{DFT}[\, x_1(n)\, x_2(n)] = \frac{1}{N} X_1(k) \circledast X_2(k)$$

The circular convolution can be interpreted as linear convolution followed by aliasing. The fact that multiplication of DFT's corresponds to a circular convolution rather than a linear convolution of the original sequences stems from the implied periodicity in the use of the DFT. The part of the signal that flows out of the period to the right, magically reappears on the left.

To handle this problem, we simply need to add each of the signals being convolved with enough zeros to allow the output signal room to handle the N+M-1 points in the correct convolution. For example, the sample signal and the impulse response could be padded with zeros to make them 512 points long, allowing the use of 512 point DFTs. After the frequency domain convolution, the output signal would consist of 306 nonzero samples, plus 206 samples with a value of zero.

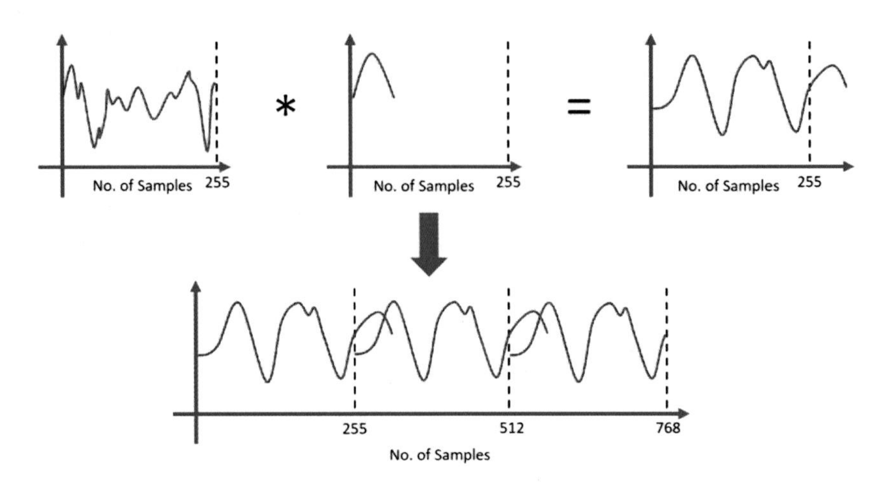

8.3 MATLAB

8.3.1 Circular Convolution

```
x = input('Enter first sequence');
y = input('Enter second sequence');
z = cconv(x,y);
subplot(3,1,1);
stem(x);
title('First input sequence');
subplot(3,1,2);
```

```
stem(y);
title('Second input sequence');
subplot(3,1,3);
stem(z);
title('Circular convolution');
```

Output:

Enter first sequence[1 2 3 4 5 6 7]
Enter second sequence[10 9 8 7 6 5]

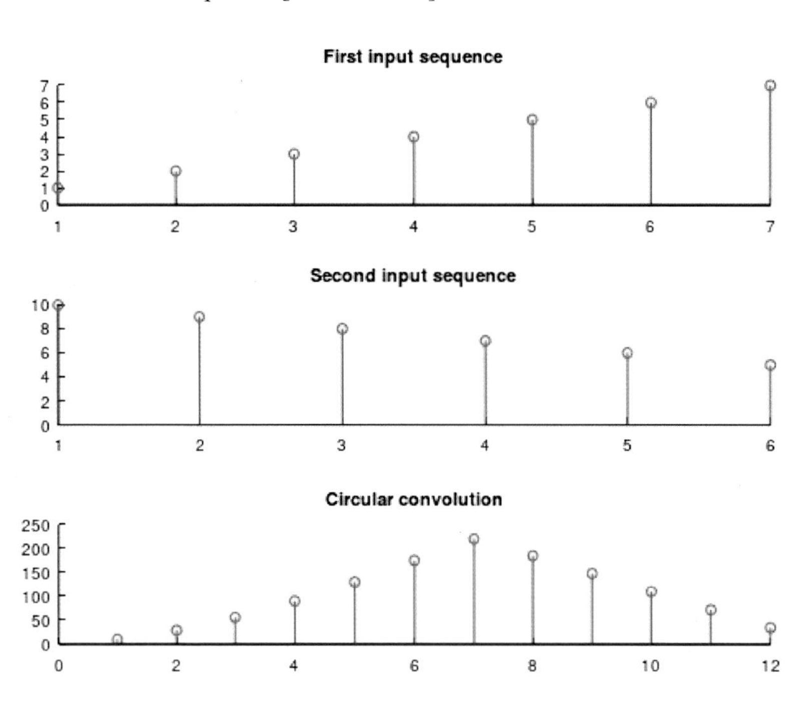

9. Z-TRANSFORM

9.1 TRANSFORMS

In chapter 3, we learnt about the Fourier Transform, and how we can obtain the frequency spectrum of an aperiodic signal using it. You might have also heard about Laplace transform from math class or in control engineering. There are other transforms too, like the Hilbert Transform, Wavelet Transform, z-transform etc.

So what exactly are transforms? Transforms are nothing more than a mathematical operation that convert a set of numbers into another set of numbers. Consider this set of numbers {1,2,3,4,5,6} .Suppose we transform this set of numbers and get another set of numbers {2,4,6,10,12}. This transform of ours converted the first set of numbers to double its value.

Now you must be thinking, if things are so simple why doesn't everyone come up their own transform and if that was the case then math books would have been flooded with transforms. Well, actually you can come up with your own transforms, no one's stopping you from doing so. If you are able to prove the usefulness of your transform or why your transform is better than existing mathematical tools, it might become a real thing. Every transform used today, are tools to make analysis easier. Some Transforms have solid intuition behind them, while some are simply created out of necessity.

9.2 Z- TRANSFORM

Mathematically, the Z-transform is defined as:

$$X(z) = Z\{x[n]\} = \sum_{n=-\infty}^{\infty} x[n]\, z^{-n}$$

Looks familiar right? Not Really? What if we told you, $\mathbf{z} = \mathbf{r}\ \mathbf{e}^{j\omega}$. Now it looks suspiciously similar to DTFT, but without **r**. If we assume **r=1**, it's exactly the DTFT. So the DTFT is actually a special case of the Z-transform.

The DTFT to the Z-transform is what the Fourier transform is to the Laplace Transform. While the Fourier and the Laplace Transform operate on Continuous time signals, the Z-transform and the DTFT operate on Discrete time signals.

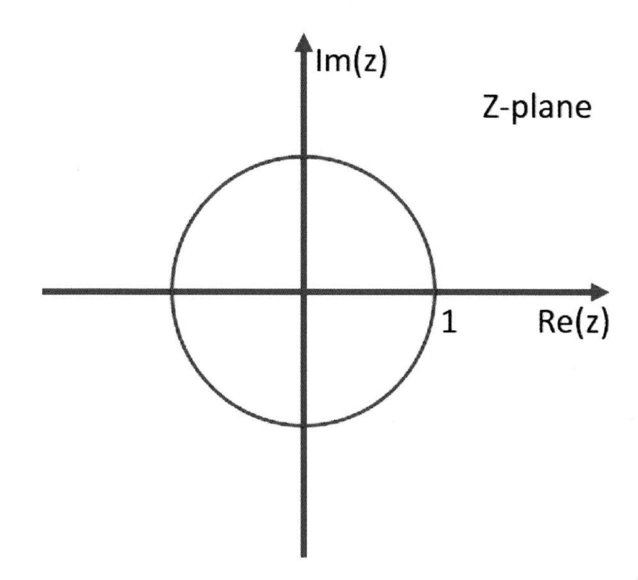

The DTFT is obtained by restricting the Z-plane to the unit circle.

	Continuous	Discrete
Aperiodic	Laplace Transform ↓ (special case) Fourier Transform	Z-Transform ↓ (special case) DTFT
Periodic	Fourier Series	DFT

So why do we need the Z-transform, when we got DTFT? The primary reason is that, the DTFT doesn't always converge/exist. That's a mathematical limitation. The DTFT exists only when:

$$\sum_{n=-\infty}^{\infty} | x[n] | < \infty$$

Whereas the Z-transform may converge where the DTFT doesn't exist. Another reason is that, using Z-transform we get a lot of polynomials, making the math a lot easier. Z-transform is widely used for stability analysis and filter design.

Having said this Z-transform also doesn't converge everywhere, it only converges in a specific region of convergence (ROC).

Remember that, Z-transform = DTFT ($\mathbf{x[n]\ r^{-n}}$). So the ROC of Z-transform can be obtained as:

$$\sum_{n=-\infty}^{\infty} | x[n]\, r^{-n}| < \infty$$

Example: $\mathbf{x[n] = a^n\ u[n]}$

This is a Right sided signal, because it exists only to the right of the origin.

$$X(z) = \sum_{n=-\infty}^{\infty} x[n]\, z^{-n} = \sum_{n=-\infty}^{\infty} a^n z^{-n}$$

$$= \sum_{n=-\infty}^{\infty} (a/z)^n$$

This function converges if **|a/z| < 1** or **|z| > |a|**

$$X(z) = \frac{z}{z-a}$$

The pole of this function is at **z=a** (value of z at which X(z) becomes ∞) and the zero of this function is at **z=0** (value of z at which X(z) becomes 0). The pole is marked as **X** and zero as **O** on the z-plane. So the ROC is very outside the circle **z=a**.

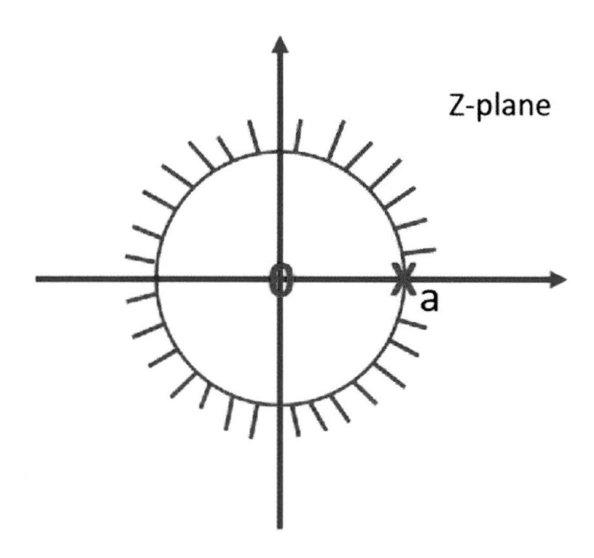

For a Left sided signal, the ROC would be inside the circle.

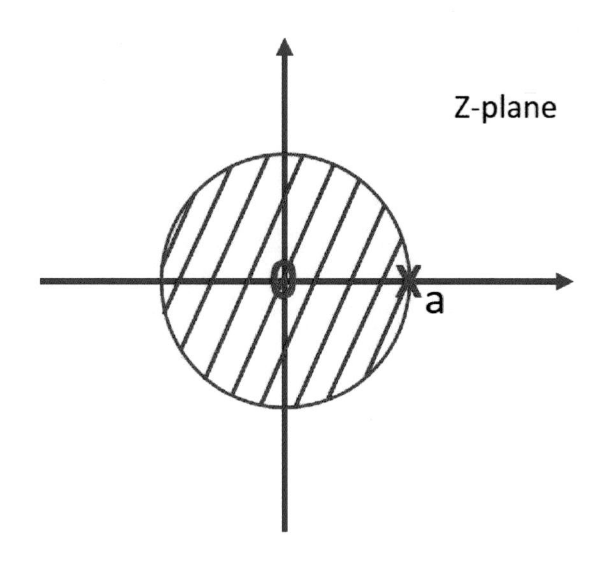

So far we have discussed about infinite length signals. For the finite length signals, things get a little easier. The ROC for a Right hand finite sequence is the entire z-plane except **z=0**. The ROC for a Left hand finite sequence is the entire z-plane except **z=∞.** For a two sided sequence (finite or infinite), the ROC will be the sum of the two conditions.

9.3 PROPERTIES OF THE Z- TRANSFORM

Many of the properties of the Z-transform are analogous to those of the Fourier transform.

9.3.1 Linearity

If $\mathbf{x_1(n)} \leftrightarrow \mathbf{X_1(z)}$ with Region of convergence $\mathbf{ROC_X}$ and $\mathbf{x_2(n)} \leftrightarrow \mathbf{X_2(z)}$ with Region of convergence $\mathbf{ROC_Y}$,

then, $\mathbf{a\,x_1(n) + b\,x_2(n)} \leftrightarrow \mathbf{a\,X_1(z) + b\,X_2(z) = W(z)}$ and the Region of convergence of $\mathbf{W(z)}$ will include the intersection of $\mathbf{ROC_X}$ and $\mathbf{ROC_Y}$ i.e. $\mathbf{ROC_W = ROC_X \cap ROC_Y}$

9.3.2 Shifting Property

If $\mathbf{x(n)} \leftrightarrow \mathbf{X(z)}$,

then, $\mathbf{x(n- n_0)} \leftrightarrow \mathbf{z^{-n0}\,X(z)}$

9.3.4 Time Reversal

If **x(n)** ↔ **X(z)**, with Region of convergence **ROC_X**,
then, **x(-n)** ↔ **$X(z^{-1})$** and Region of convergence of **$X(z^{-1})$** will be **ROC_Y = 1/ROC_X**

9.3.5 Conjugation

If **x(n)** ↔ **X(z)**, with Region of convergence **ROC_X**, then the Z-transform of the complex conjugate of **x(n)** is
x*(n) ↔ **X*(z*)** and Region of convergence remains the same

9.3.6 Derivative of X(z)

If **x(n)** ↔ **X(z)**, then the z-transform of **n x(n)** is
n x(n) ↔ **z d(X(z))/dz**

9.4 CONVOLUTION THEOREM AND STABILITY

If **x(n)** ↔ **X(z)** with Region of convergence **ROC_X** and **h(n)** ↔ **H(z)** with Region of convergence **ROC_Y**, then **x(n) * h(n)** ↔ **X(z) H(z)** = **Y(z)**, and the Region of convergence of **Y(z)** will include the intersection of **ROC_X** and **ROC_H** i.e. **$ROC_Y = ROC_X \cap ROC_H$**. Here **h(n)** denote the unit impulse response and the Z-transform of unit impulse response is denoted by **H(z)**.

For an LTI system, the condition for BIBO stability is,

$$\sum_{n=0}^{\infty} |h(n)| < \infty$$

For a causal sequence,

$$H(z) = \sum_{n=0}^{\infty} h(n)\, z^{-n}$$

$$|H(z)| = \sum_{n=0}^{\infty} |h(n)|\, |z^{-n}|$$

So $|H(z)|$ satisfies the condition for BIBO stability, when $|z| = 1$, i.e. a causal LTI is BIBO stable, if and only if all the poles of **H(z)** are inside the unit circle.

10. FILTERS

10.1 INTRODUCTION TO FILTERS

In signal processing, the function of a filter is to remove unwanted parts of a signal and to extract the useful parts. There are two main kinds of filters: analog and digital. They are quite different in their physical makeup and in how they work.

An analog filter uses analog electronic circuits made up from components such as resistors, capacitors and op amps to produce the required filtering effect. Such filter circuits are widely used in applications such as noise reduction, video signal enhancement etc. There are well-established standard techniques for designing an analog filter circuit for a given requirement. At all stages, the signal being filtered is an electrical voltage or current which is the direct analogue of the physical quantity (e.g. a sound or video signal or transducer output) involved.

A digital filter on the other hand uses a digital processor to perform numerical calculations on sampled values of the signal. The processor may be a general-purpose computer such as a PC, or any other processor. The analog input signal must first be sampled and digitized using an ADC (analog to digital converter). The resulting binary numbers representing successive sampled values of the input signal are transferred to the processor, which carries out numerical calculations on them. Then the results of these calculations, which now represent sampled values of the filtered signal are output through a DAC (digital to analog converter) to convert the signal back to analog form.

Note that in a digital filter, the signal is represented by a sequence of numbers, rather than a voltage or current.

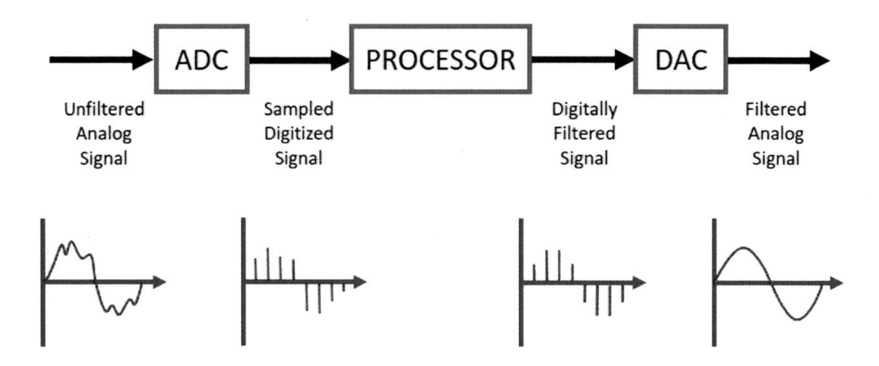

10.2 ADVANTAGES OF DIGITAL FILTERS

Digital Filtering is one of the most powerful applications of Digital signal processing. In fact the popularity of DSP can be primarily attributed the success of digital filters. Digital filters have some major advantages over Analog filters:

1. Digital filters allow us the luxury of changing filter characteristics by merely changing some lines of codes. This is a challenging part for their analog counter parts, as it requires hardware changes.

2. It is easier to design, test and implement a Digital filter.

3. Digital filters aren't effected by hardware factors.

4. Digital filters are more versatile. Any processor can function as digital filter by merely transferring some software.

10.3 DIGITAL FILTERS

Technically speaking, a filter is nothing more than an LTI system, so all the properties, impulse response, frequency response etc. are applicable to it.

The most straightforward way to implement a digital filter is by convolving the unfiltered input signal with the filter's impulse response (called filter kernel). All possible linear filters can be made in this manner. Filters made this way are called FIR or Finite Impulse Response Filters.

There is another way to make digital filters called Recursion (means running back). This is an extension to the previous method. In this method, previously calculated values from the output stage is used, besides the input values. For Recursive filters, their impulse responses are infinitely long. Hence they are also called IIR or Infinite Impulse Response Filters.

From this explanation, it might seem as though IIR filters require more calculations to be performed, since there are previous output terms in the filter expression as well as input terms. In fact, the opposite is usually the case.

Consider this simple example of an IIR filter: $y_n = x_n + y_{n-1}$

This filter determines the current output (y_n) by adding the current input (x_n) to the previous output (y_{n-1}):

$y_0 = x_0 + y_{-1}$,

$y_1 = x_1 + y_0$,

$y_2 = x_2 + y_1$ .... etc.

To achieve the same with an FIR filter, we would need to use the expression:

$$y_n = x_n + x_{n-1} + x_{n-2} + \ldots + x_1 + x_0$$

From this example it's very evident that using IIR filters, the process would become faster (lesser operations to do).

The general expression for any digital filter is,

$$y(n) = -\sum_{k=1}^{N} a_k\, y(n-k) + \sum_{k=0}^{M} b_k\, x(n-k)$$

Taking Z-transform on both sides, the transfer function of a filter can be obtained as:

$$H(z) = \frac{Y(z)}{X(z)} = \frac{\sum_{k=0}^{M} b_k z^{-k}}{1+\sum_{k=1}^{N} a_k z^{-k}}$$

For FIR filters, $\{a_k\} = 0$, since feedback is absent.

The order of a digital filter is defined as the number of previous input or output values required to compute the current output. The filter in our example is a first order filter, because it uses one previous output (y_{n-1}) to compute present output value (y_n).

Example:

A filter with transfer function $H(z) = 1 + 2z^{-1} + 4z^{-2}$, can be implemented as shown in the figure below.

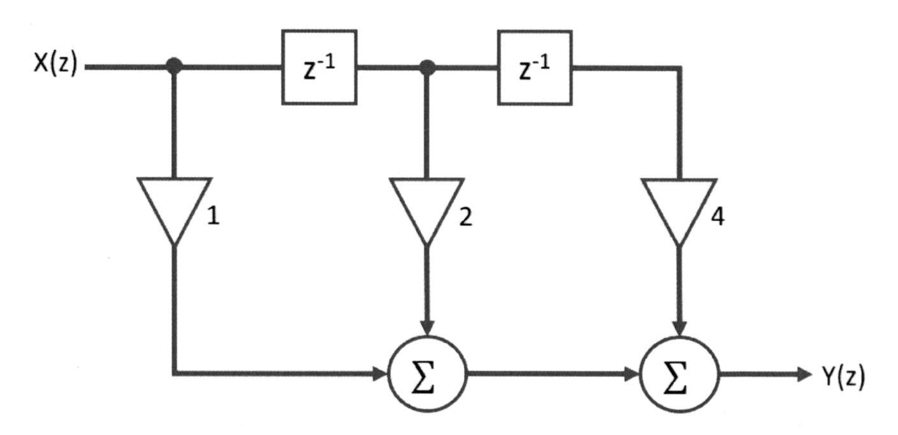

While designing digital filters, various factors have like stability, order etc. has to be considered for real life implementation. Filter design is a vast topic, requiring a heavy dose of mathematics, which is beyond the scope of this book. Nonetheless, the basic ideas regarding filters and LTI systems in general has been discussed already.

APPENDIX

1. EULER'S FORMULA

Euler's formula, named after famous mathematician Leonhard Euler, is a mathematical formula that establishes the fundamental relationship between the trigonometric functions and the complex exponential function.

$$e^{i\phi} = \cos\phi + i\sin\phi$$

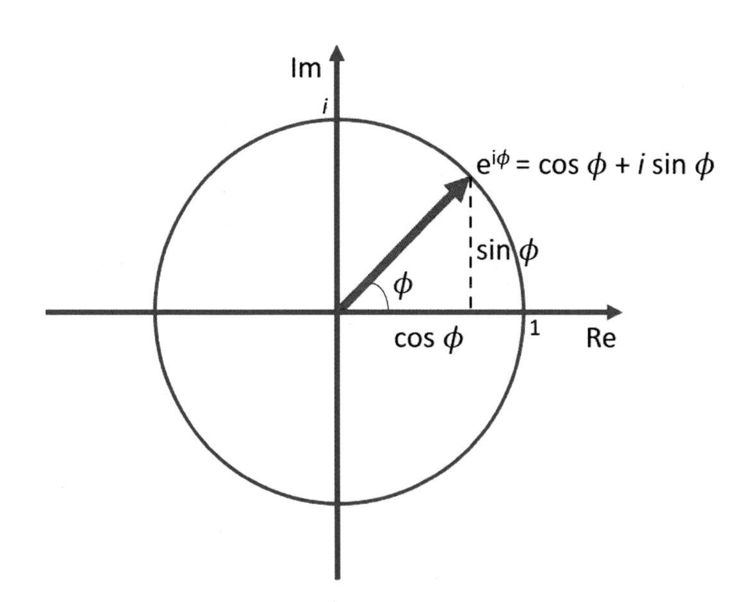

This formula can be interpreted as saying that the function $e^{i\phi}$ is a complex number of unit magnitude, for all real values of ϕ. Euler's formula provides a means of conversion between cartesian coordinates and polar coordinates. Euler's formula also provides an interpretation of the sine and cosine functions as weighted sums of the exponential function.

2. FOURIER REPRESENTATIONS

	Periodic	Aperiodic
Continuous signals	Fourier Series $X[k] = \frac{1}{T} \int_{0}^{T} x(t)\, e^{-jk\omega t}\, dt$ Inverse Fourier Series $x(t) = \sum_{k=-\infty}^{\infty} x[k]\, e^{jk\omega t}$	Fourier Transform $X(\omega) = \int_{-\infty}^{\infty} x(t)\, e^{-j\omega t}\, dt$ Inverse FT $x(t) = \frac{1}{2\pi} \int_{\omega=-\infty}^{\infty} X(\omega)\, e^{j\omega t}\, d\omega$
Discrete signals	DFT $X[k] = \sum_{n=0}^{N-1} x[n]\, e^{-jk(2\pi/N)n}$ Inverse DFT $x[n] = 1/N \sum_{k=0}^{N-1} X[k]\, e^{jk(2\pi/N)n}$	DTFT $X(\omega) = \sum_{n=-\infty}^{\infty} x[n]\, e^{-j\omega n}\, d\omega$ Inverse DTFT $x[n] = \frac{1}{2\pi} \int_{\omega=-\pi}^{\pi} X(\omega)\, e^{j\omega t}\, d\omega$

REFERENCES

1. **Advanced Engineering Mathematics** by Erwin Kreyszig
2. **Digital Signal Processing: Principles, Algorithms, and Applications** by J. G. Proakis and D. G. Manolakis.
3. **Discrete-Time Signal Processing** by A. V. Oppenheim and R. W. Schafer.
4. **Signals and systems** by Alan V. Oppenheim
5. **The Scientist and Engineer's and Guide to Digital Signal Processing** by Steven W. Smith.
6. **Understanding Digital Signal Processing** by Richard G. Lyons.

OTHER BOOKS IN THE SERIES

Arduino for Complete Idiots

Control Systems for Complete Idiots

Circuit Analysis for Complete Idiots

Visit us at BeyondWhyy.com

Made in United States
North Haven, CT
29 December 2022